Thank-you...

for ordering from Hello Direct! Please accept this informative book as our gift to you. You can use this book many ways: for business or personal use, or as a gift to colleagues, friends or family.

We want you to be completely satisfied with everything you order from Hello Direct. Therefore, if you have any questions or concerns about *anything* you've received from us, please call 800/444-3556. Remember, our trained support staff is here to help you in any way we can.

Once again thank-you, we value your business. And if you need anything else, just call 800/444-3556,

The Hello Direct Team

What They're Saying About *Telecom Made Easy*:

"Suddenly, the small-office/home-office market is hot, spawning a number of new how-to books geared toward the soloist. *Telecom Made Easy* is one we like for its thorough coverage of telephones for home and small businesses. . . Especially helpful for do-it-yourselfers. . . Just leafing through this 380-page book will prepare anyone for smarter dealings with salespeople or the phone company."

— *INC.* magazine

". . . a basic but thorough guide to phone systems and services, cellular phones, answering devices, paging, on-line services, modems, faxes, and networked systems. . . geared toward home businesses, telecommuters, and small firms."

— *Nation's Business*

". . . offers lots of money-saving phone system solutions for small businesses."

— *Independent Business*

". . . how to streamline a company and its phone use, and how to provide better services through the phone. An excellent, basic guide."

— *The Midwest Book Review*

"I used to call Langhoff when I was writing my column and had a question about telecommunications and the home office. Now I can just pick up this book. This is a great resource."

— Alice Bredin
nationally syndicated *Working at Home* columnist

"My dog-eared copy has yet to see my bookshelf."

— Mark Anthony, President
Home Based Business Assn. of Minnesota

Other Telecom Titles from Aegis Publishing Group:

Telecom Business Opportunities
The Entrepreneur's Guide to Making Money in the Telecommunications Revolution, by Steve Rosenbush
$24.95   1-890154-04-0

Winning Communications Strategies
How Small Businesses Master Cutting-Edge Technology to Stay Competitive, Provide Better Service and Make More Money, by Jeffrey Kagan
$14.95   0-9632790-8-4

Telecom & Networking Glossary
Understanding Telecommunications Technology
$9.95   1-890154-09-4

Telecom Made Easy
Money-Saving, Profit-Building Solutions for Home Businesses, Telecommuters and Small Organizations, by June Langhoff
$19.95   0-9632790-7-6

The Telecommuter's Advisor
Working in the Fast Lane, by June Langhoff
$14.95   0-9632790-5-X

900 Know-How
How to Succeed With Your Own 900 Number Business, by Robert Mastin
$19.95   0-9632790-3-3

The Business Traveler's Survival Guide
How to Get Work Done While on the Road, by June Langhoff
$9.95   1-890154-03-2

Getting the Most From Your Yellow Pages Advertising
Maximum Profits at Minimum Cost, by Barry Maher
$19.95   1-890154-05-9

Money-Making 900 Numbers
How Entrepreneurs Use the Telephone to Sell Information
by Carol Morse Ginsburg and Robert Mastin
$19.95   0-9632790-1-7

How to Buy the Best Phone System
Getting Maximum Value Without Spending a Fortune, by Sondra Liburd Jordan
$9.95   1-890154-06-7

1-800-Courtesy
Connecting With a Winning Telephone Image, by Terry Wildemann
$9.95   1-890154-07-5

The Telecommunication Relay Service (TRS) Handbook
Empowering the Hearing and Speech Impaired, by Franklin H. Silverman, Ph.D.
$9.95   1-890154-08-3

TELECOM MADE EASY

Money-Saving, Profit-Building Solutions
for Home Businesses, Telecommuters
and Small Organizations

Third edition

by June Langhoff

Aegis Publishing Group, Ltd.
796 Aquidneck Ave.
Newport, RI 02842
401-849-4200

illustrated by Vicki Zimmerman

Library of Congress Catalog Card Number: 97-72689

Aegis Publishing Group, Ltd
796 Aquidneck Ave., Newport, RI 02842

International Standard Book Number: 0-9632790-7-6

Printed in the United States of America.
First Edition, 1995. Second Edition, 1996. Third Edition, 1997.

10 9 8 7 6 5

Publisher's Cataloging in Publication Data
Langhoff, June
Telecom Made Easy: Money-Saving, Profit-Building Solutions for Home Businesses, Telecommuters and Small Organizations / by June Langhoff—Third edition
Includes Index
1. Telecommunications.
TK5101.L35   1997   621.382   97-72689
ISBN 0-9632790-7-6

Acknowledgments

●●

Some super people helped put this book together and I'm most grateful for their help. My heart-felt appreciation to:

Bob Mastin, my editor and publisher, for his unfailing support and enthusiasm for this project.

Patrick Rentsch, for patiently answering hundreds of technical questions and wading through very rough drafts.

Nim Marsh, for copy-editing the manuscript with great delicacy and care.

Vicki Zimmerman, for the exceptional illustrations that pepper the book and help explain technical stuff.

John Robertson, for an outstanding cover design.

Kim Ecclesine, for cheerfully snapping my photo for the back cover.

David Joseph Lacagnina, for access to Salestar, his company's telecommunications database.

My friends at Pacific Bell for all their support and advice.

The folks that hang out on the Working From Home Forum on CompuServe, who willingly answered my questions and signed up for many of the interviews that are sprinkled throughout the book.

My office mates—Max, Winston, Frisby and Marsha—for keeping their muddy paws off the manuscript.

And finally, thanks to my terrific son Nick, for last-minute proofing and for putting up with this mess for months.

...Contents

Foreword

● ●

"Mr. Watson, come here. I want you."
—*Alexander Graham Bell to Thomas Watson, March 10, 1876*

As an entrepreneur living in an era of instant information, I often wonder what Alexander Graham Bell would think of modern telecommunications. What remarkable changes have occurred since Bell spoke those historic first words!

Imagine Bell's office if he were alive today. Multiple telephone lines would connect him to colleagues around the world by voice, fax, and modem. He'd probably be a master surfer of the Internet, gathering data for new inventions from points around the globe. On his busy travels, he'd stay in touch by cellular phone, having his messages routed from his voice mail system or pager. His laptop computer would be a constant companion, since he'd be using it to exchange e-mail, data files, faxes, and audio and video clips from remote sites as he visited clients around the world. Interested in learning more about his business? You could call his office's toll-free 800 number for general information, use his pay-per-call 900 number to hear valuable details on recent developments, or access his fax-on-demand system for printed reports of his new patents and inventions.

While this imaginary scene may seem a fanciful depiction, it is the everyday realm of millions of 1990s entrepreneurs who employ these technologies on a regular basis in their work. If Bell were alive today, however, it's also likely that he'd be surrounded by fellow entrepreneurs eager to discover how to use telecommunications effectively in their own businesses—and besieging him to help them decipher the confusing array of options and information.

Since Alexander Graham Bell can be our guide only in our imagination, we are fortunate to have June Langhoff's expertise in this book to assist us with our real-world needs. In clear, easy-to-understand language, she demystifies the complexities of this powerful technology. Her straightforward approach is the perfect fit for busy entrepreneurs who are more interested in running their businesses than becoming telecommunication engineers.

Savvy entrepreneurs understand that staying connected is a key ingredient to success—and that telecommunications is at the heart of those connections. Yet who among us has not been overwhelmed at the fast-paced changes and multiple decisions to make regarding this area? Modern technology enables us to appear like and compete with businesses many times our size. Unlike those larger companies, however, we don't have a dedicated staff member to manage this technology. The potential it brings for empowering our business is thrilling, but the details can make us dizzy.

June Langhoff is here to guide us. Consider her your personal telecommunications consultant—at a fraction of the cost! Never before has a single book provided such a valuable overview of the telecommunications information small business owners need to make their businesses succeed. Use it to find the answers you seek today, and keep it close at hand to tackle those telecom challenges you'll face in the future.

More than any other development, advances in telecommunications have fueled the explosive growth of entrepreneurial activity in our country. Recent studies show that more than 41 million Americans are involved in solo businesses, home-based enterprises, and telecommuting arrangements. Booming numbers, how-ever, do not guarantee success for all. To compete successfully, today's entrepreneurs must adapt technology for their own uses and harness it effectively. And to master technology, you must first *understand* it.

This book gives that power. Let's cheer June Langhoff for bringing us real-world solutions to high-tech confusion. May the information and advice she has gathered here bring you much success.

Terri Lonier
New Paltz, New York

Professional speaker and consultant Terri Lonier is president of Portico Press, publishers of books, newsletters, an online network, audiotapes, and other resources for entrepreneurs. She is the author of *Working Solo*, the complete guide to self-employment chosen by *Inc.* magazine as the number-one book for solo entrepreneurs and one of the top three small business books in the U.S.

As a 1990s entrepreneur, Terri reports that she runs her business with four telephone lines, voice mail, a dedicated fax line, a cordless phone, an answering service, four online addresses, and several modem-equipped computers. For information on her resources for entrepreneurs, call the Portico Press toll-free number (800-222-SOLO) and ask for a free brochure.

1...Communications Today

Telephones ... *yawn*. Telecommunications ... *borrrring*. If that's how you feel about the subject of telecom, it's time for you to wake up.

No business today, no matter how small, can afford to remain ignorant about telecommunications. If you want to compete with large, resource-laden organizations, you have no choice but to sound like one yourself. You do that by creating a communications system that handles phone calls, messages, fax, and data transfer with ease. One that keeps you instantaneously available.

That's not easy. But it's getting easier. How? Through employing telecommunications products and services such as those outlined in this book.

Look at what is happening in the communications world today:

- Avid windsurfers along the California and Hawaii coastlines get the latest wind conditions reported on their Call of the Wind pagers. An automatic monitoring system alerts customers when the wind is right.

- A local SuperCuts in Aurora, Colorado operates a completely equipped hair salon in a RV. In addition to the usual stuff a beautician would carry, the stylist also packs a cellphone (for last-minute appointments) and a cellular credit-card authorization machine.

- Residents of Keene, New Hampshire, can call the local humane society, select a category from the audiotext menu (cat, dog, bunny), and listen to pre-recorded descriptions of animals available for adoption.

- At the High Tech Cafe in Dallas, the maitre d' asks diners if they prefer "smoking, nonsmoking or modem-ready." Tables at the Cafe are equipped with modem jacks and power cords so that patrons can grab a fax, fire off an e-message, continue computing, or charge their batteries as they do lunch.

- The sheriff's office in Cupertino, California is part of City-Net, a community network. Citizens can go online for an electronic chat with law enforcement officers, ask questions about public safety, or view a video database of photos of missing children.

- Internal paper mail (p-mail) is banned at VeriFone, a credit card authorization company. All employees are required to use electronic mail instead. Everything about the company is online—shipments, project tracking, travel schedules, and forms.

- When wining or dining at a local TGIF, patrons request service by speaking into their saltshaker. A pager, hidden inside the condiment holder, summons the waiter.

- Baynet, an online network, offers a useful service to the real estate industry. For a small fee, it will scan a color photo and flyer of a listed property and upload the image on the World Wide Web. That listing can be accessed by over 30 million prospective buyers.

- Farmers are equipping their tractors with cellular phones. Then, if they have a breakdown, they can dial a repair company directly from the field, saving a trip to town. Some are even putting pagers on their dairy cattle, summoning them home with a beep.

- Filmmaker Steven Spielberg edited *Jurassic Park* while working in Poland on his next movie using a broadcast video service from Pacific Bell. The system, dubbed "Virtual Hollywood," employs satellite, microwave and fiber technologies to move huge amounts of digitized film data.

Our world is shrinking. Telecommunications is changing the way we work, the way we communicate. Organizations are doing business in new forms. Online ordering, cybermalls, virtual organizations—these are the business innovations of the last decade of the 20th century. Be a part of this telecommunications revolution. Avoid being left at the side of the road, like road kill on the info-highway.

A 1994 Coopers & Lybrand survey of the fastest-growing small businesses found that revenue per employee is two and a half times higher for companies that use high technology compared to low-tech companies. The mere fact that you're reading this book most likely places you in the former category. Or at least wanting to get there. Good for you!

Why join the telecom revolution?

You can use telecommunications to:

✔ Provide great customer service

Set up an 800 number for customers to call. Create a fax-back service so they can get instant help.

✔ Stay available

Carry a voice pager, PCS phone, cellphone or radio modem to keep in touch. Sign up for follow-me phone service or call forwarding.

✔ Discover new ways to market

Create a newsletter and use a fax modem to send it to your client list. Advertise online. Network in cyberspace.

✔ Speed workflow

Add a network. Use workgroup software, voice broadcast, and e-mail distribution lists to keep everyone focused. Install a high-speed digital phone line for faster faxing and accelerated file transfer.

✔ Tap into instant information

Subscribe to online databases that can help you in your work. There are specialized databases and BBSs for lawyers, engineers, teachers, investigators, veterinarians, to name a few.

✔ Save money

Install a telephone system for maximum line usage and for controlling long-distance costs. Telecommute instead of traveling. Invest in a faster modem.

✔ Add to revenues

Get a toll-free number and watch orders increase. Set up a 900 number or fax-back service to sell information and services.

✔ Improve security

Install an intercom system from your desk to your door. Take a cellular phone with you on the road.

✔ Save time

Avoid centralized meetings by scheduling a teleconference or setting up a desktop video system. Take advantage of broadcast fax services.

✔ Expand your market

Set up a remote line to snare business from another community. Use phone company voice mail or targeted 800 or 888 numbers to widen your cast.

✔ Stay connected

Schedule team meetings online. Use document conferencing and remote control to smooth out the bumps in the collaborative process.

✔ Work smart

Carry your office with you in a briefcase, backpack or bag. Use e-mail or voice mail to communicate with far-flung colleagues and customers.

When I was conducting research for this book, I found several books that tell you how to *save* money on telephone service, but I didn't find any that advised how to *make* money using telecommunications. That's what I plan to do. In order to make money using telecommunications, you need to think of your communications budget as a marketing expense, not as a fixed overhead expense.

Think back to 1977. That was the year that the first personal computers appeared on the market. Those chunky Apple computers looked like a cute fad, didn't they? Then IBM came out with a personal computer, and the *fad* was taken much more seriously. Some businesses embraced the technology, found new niches for themselves, and took off. Witness desktop publishing, multimedia, and online shopping services. Other businesses took a wait-and-see attitude. Some of them are still around. Many are not.

We're approaching a similar crossroads today in the arena of telecommunications. Organizations that are taking advantage of new developments in telecom are increasing their chances for survival tomorrow. Plan to be one of them.

Happily, you don't have to be a large organization with megabucks to spend in order to join the telecom revolution. Most of the new technologies are well within the budget of small and solo businesses. For example, you can have voice mail for about $6 a month, a personal toll-free number for another $5, and e-mail paging starts at around $30. You can add an Internet account for $10 to $20 a month.

Be glad you're small. Your lean size lets you make quick decisions, implement rapid changes, and embrace new technology much sooner than the behemoths. Though a couple hundred dollars spent on a speedy modem may seem like a lot to you, multiply that by thousands and think about how much longer it takes large organizations to make technology decisions. New technologies are often easier to implement in a small organization. You're not trammeled by layers of management, acres of paperwork, and tangled lines of communication.

Evaluating your needs

Do your communications match your customer's expectations? If your customers are used to doing business by fax, you're losing business by not having fax capability. If they are dedicated e-mailers, make sure you have an e-mail address of your own. If your clients expect to reach you at a moment's notice, figure out how to accommodate them via pager, cellular, or call forwarding. Upgrade your image by adding a dedicated line for fax and modem calls and get a voice mail system.

Have you ever heard one of your customers, vendors or clients tell you that "your office is very hard to reach"? Probably not. They usually don't tell you when they decide to call someone else. It's time to take control of your communications.

When deciding where to invest your telecom dollars, you need to start with a plan. What do you want to accomplish? Increased accessibility? Wider market share? Speedier ordering? Improved customer service? Maybe your current push is to reduce costs, do things smarter.

Each day communications choices get more complex. If you look in the newspaper or scan computer and telecom magazines, you're hit with an incredible selection of products, all clamoring for your attention. Each having a longer feature list than the last. Full of indecipherable technotalk. How do you sort it all out?

First, write down all your questions. Yes, all of them.

- Do I need a stand-alone fax? A fax/modem board? Or a fax/voice/modem card? Should it be wired or wireless? Will it work with the snazzy digital key system I just bought last year?

- Maybe I should wait before buying that cellular phone? It looks like there will be a lot of change in the coming months. Or could I avoid getting a cellular phone and just use an alpha pager or PDA with e-mail?

- It seems as though everyone in the world has voice mail except me. Should I get it, too? Or maybe I should get a multifunction device that takes voice, fax and e-mail messages?

- I really want to be a node on the road, but how? Is there any way to hook up my computer at home with my computer at work? If there is, will my company let me do it? If I leave my computer on all the time, will I be open to security breaches, hackers and worse?

Next, do your homework. Read this book (well, the applicable chapter anyway). Once you know what all those features in the ads mean, make a list of the features you want. If the list is long, prioritize it.

Now you're ready to take a trip to a local electronics or telecom store. Make a vow not to buy on this trip. It may be hard to resist the sales pitch, but you need to learn what's available and how everything works before making a decision. Try stuff out. Ask for demos to see how the features work.

Once you're safely back in your office again, call around to compare prices and features. Consider buying used equipment. Take advantage of leasing plans. Check out the mail-order catalogs.

Do the numbers. What's the payback period? What's the useful life of the product? What are the operating costs? Can you get a multifunction device that handles several of your needs?

Finally, make your decision and then implement it. Don't wait for the next round of technological improvements. If you do, you'll never be able to join the telecom revolution. You'll just sit on the sidelines, salivating over technical reviews, and suffering from acute PBX, PDA or PCphone envy.

How to use this book

Skim this book. Look at the introductions to each chapter. Take the quick surveys—designed to help you select technologies—that are scattered throughout the book.

You certainly don't have to read every word or every chapter to get the information you need. Both the Table of Contents and the Index can point you to areas of interest.

To help you get set up easily, most chapters have a section on getting connected. Often, I've included feature lists, shopping advice and troubleshooting tips. At the end of every chapter, you'll find a list of resources—magazines, books, newsletters, online sources, web sites, vendor names and phone numbers—hopefully, everything you'll need to get additional information on a particular topic or product.

Here's a brief rundown of each chapter's contents:

Chapter 1 - Communications Today

You're here already. But, in case you missed it, this chapter piques your curiosity about the telecom world by showcasing new telecommunications technology. I tell what's in it for you, provide some advice on determining your needs and give you an overview of each chapter.

Chapter 2 - Connections

This is the "nuts and bolts" of telecommunications—including wiring plans, connections, tips for getting the most from your current setup, and some troubleshooting advice. There's even a small section for the do-it-yourselfer. Be sure to read this chapter if you plan to add more phone lines to your phone system. What you do now could save bundles in the future.

Chapter 3 - Phone Services

Dozens of telephone company services that can improve your communications are listed. You'll learn about Caller ID, Follow-me-Phone Service, Distinctive Ringing, Call Waiting and all kinds of call forwarding.

Chapter 4 - Phone Lines

You'll get the low-down on money-making services, such as 800 and 900 lines. I explain the difference between analog and digital phone lines, describe line service types such as POTS, ISDN, and T-1 in detail, and provide tips on ordering phone service.

Chapter 5 - Stand-Alone Phones

This chapter is all about phones—how they work, what all those different features do, and what to look for in a single-line, multi-line or cordless phone. There's advice about troubleshooting and information about selecting and using a headset. Finally, there's a section on new telephone technology, including phones that operate through your personal computer.

Chapter 6 - Phone Systems

You'll learn all you need to know to decide what kind of phone system you need (and if you even need one at all). I discuss key, KSU-less, PBX, Centrex, hybrid and PC-based phone systems. You'll learn what all those features mean and which ones are most useful to you.

Chapter 7 - Mobile Phones

Mobile phones—including portable, transportable and car phones—are the topic of this chapter. You'll get advice on selecting a mobile phone and carrier, useful information on keeping your cellular bill under control, and learn the latest technological advances such as PCS and personal satellite communications.

Chapter 8 - Voice Mail

You'll need a messaging system to keep track of your incoming calls. This chapter explains voice mail systems and voice mail services. Included are useful tips for setting up user-friendly voice mail, designing a successful voice mail menu and tips for leaving effective messages.

Chapter 9 - Answering Machines

Answering machines are becoming much more than just mechanical message takers. This chapter explains new answering machine technology, including digital answering devices, portable answering, and multifunction machines.

Chapter 10 - Paging

Pagers aren't low-tech beepers anymore. This chapter discusses various types of pagers, explains pager features, and provides advice on shopping for a paging service. New technologies such as PC-based paging and radio modems are also explained.

Chapter 11 - Going Online

Use this chapter as your on-ramp to the information highway. We start with an overview of online services, how to sign up for an online service, and how to get on the Internet. There are special sections on using e-mail, transferring files, and some tips on how to protect against computer viruses. I also provide advice on how to do business on the Internet.

Chapter 12 - Modems

If you want to explore the online world, fax from your computer, or participate in electronic data interchange networks, you'll need a modem. This chapter simplifies the task of choosing a modem, tells you what all those blinking modem lights mean, and takes the mystery out of words like bandwidth and baud. You'll also find trouble-shooting tips, directions for getting connected and advice for using your modem internationally.

Chapter 13 - Fax

Do you need a fax machine? Or maybe you want to get a fax modem for your computer? This chapter spells it all out, compares the various types of fax technologies, and helps you make up your mind. Since many small businesses want to save telephone expenses, there are tips on sharing a line and getting connected.

Chapter 14 - Telecommuting

This chapter is for any home office worker or small businessperson who wants to work remotely. I've included tips and tools for working from home and for working on the road. A special section examines several technologies that aid telework such as document conferencing, remote control software, and video conferencing.

Chapter 15 - Your Phone Bill

Staying on top of telephone costs and sorting through all those conflicting, confusing long-distance claims can leave you reeling. This chapter helps you get a handle on controlling phone costs and selecting the best calling plan for you.

Chapter 16 - Future Planning

Not all small businesses stay small. This chapter examines some approaches that aid growth such as local area networking and electronic data interchange. I'll also provide advice about disaster-proofing your business and discuss how many lines you should have. Finally, I'll profile some actual businesses and show how they handle their communications needs.

Appendices

You might want to dip into the appendices for additional tele-communications books, magazines and websites. There's also a list of major telco companies in the United States and Canada, a short glossary of online lingo, and a comprehensive index.

Things change fast around here

From the beginning of this project, I wanted *Telecom Made Easy* to be a really useful sourcebook, with names, telephone numbers and even price ranges. I wanted it to contain the information you'd need to take action, get more information, make an informed decision. I believe the book achieves these goals.

However, as you know, the telecom world is rapidly changing. Companies merge, change their names, add and drop products, revise price lists, get born, die. So I can't promise that every telephone number listed is still in service or that every product described is still available. I can promise that the information contained is as up-to-date as I could make it.

Let's keep in touch

I'm very interested in your feedback and would like to hear from you. What was useful to you? What was not? Maybe you have a story you'd like to see in the next edition or a product you'd like to see featured. If you want to contact me, please write in care of:

Aegis Publishing Group
796 Aquidneck Avenue
Newport, Rhode Island 02842

You can also reach me directly via e-mail at 71022.2131@com-puserve.com

2...Connections

● ●

Now that the telephone companies have been restructured, you are responsible for the installation, repair and maintenance of your own telephone service. Actually, you are responsible for your interior wiring (the phone lines that run inside your building), and the phone company is responsible for exterior wiring (everything else). You can do the wiring yourself, pay a electrician or contractor to do it, or pay the telephone company to do it.

If you think you'll need to troubleshoot your communications setup, or add capacity, it's a good idea to understand how telephone communications work. At the very least, you want to find out where your responsibility ends. Knowing a few basic facts about your phone system could prevent the unhappy possibility of paying a whopping service fee to your phone company, only to find that the problem is caused by a fault in your interior wiring, a shorted phone, or by a flaky answering device.

This subject can be intimidating, but, believe me, if I can understand this stuff, you can too. I promise to try to avoid complex electrical wiring diagrams and technical talk. I also promise to tell you only what you need to know and why you need to know it.

The anatomy of a phone call

- **Pick up receiver:** When you pick up the phone receiver, a spring-loaded button on the phone cradle pops up. This allows current to flow into your telephone and notifies your telephone central office that you are requesting a line. The central office finds an unused circuit and sends dial tone down the line.

- **Dial the number:** Once you have keyed in the first three numbers, your central office switch will search for a circuit to connect you to the central office of the number you are dialing. It will send a ringing tone to you to let you know that your call is being processed. Meanwhile, it will start ringing the phone of the number you are dialing.

- **Receiving party picks up:** When the person at the other end of the call picks up the phone, current flows from that phone line to their central office. This tells the central office to stop ringing the phone and to set up the audio connection.

- **At the end of the call:** When the call is completed, you hang up the phone by placing the receiver back in the cradle. This causes the button in the phone cradle to be depressed and sends a signal to the central office to release the circuit and make a record of the call's duration.

Long-distance call routing

If the call is a long-distance call, the central office switch will recognize it by the number "1" preceding the country code and/or area code. The switch contacts an *interexchange carrier* (the company you purchase your long-distance service from) and searches for a *trunk* to carry the call. Your call may actually be routed over several switches on its way. Most of the time, you won't even notice this because it happens so fast.

Calls that must cross an ocean or other large body of water will pass their signals either from a *ground station* to a communications *satellite* or over an *undersea cable*. Some calls may also travel via *microwave relay towers*.

Your connection

Telephone service is brought to your home or office via underground or overhead cables from the phone company's central switching office. (For more about central offices, see *Your Local Central Office* in Chapter 4.) The line that connects your location to the telephone company is called "the local loop."

Phone service is brought to a box that can usually be found outside your building or in the basement. This box contains a grounding device (called the station protector) to protect against lightning. It also contains some circuitry for testing. The box goes by many different names: access box, service box, station protector, junction box, demarcation block.

Station Protector

If your phone wiring was done after 1986, you'll most likely find a phone jack inside the access box. This jack, called the network interface jack, is used for testing (more about this later in this chapter). This is the point where the phone company's responsibility for your service ends (unless you pay a monthly fee for inside wiring maintenance).

In multi-unit buildings, phone service is brought to the outside of the building and comes to individual suites or apartments through risers, a bundle of wires often enclosed in a thick cable or grouped inside a raceway. Individual telephone wires are then "peeled off" to supply telephone service to office suites or apartments. There is usually one access box for each floor. This can often be found inside a utility closet or stairwell. It then is your responsibility to pay for wiring and maintenance from the general access box on your floor to your office. In a large building, this can be a substantial distance.

If your office has multiple lines your service is probably wired to a punchdown-block. The phone company calls this the distribution frame. This device looks somewhat like a bunch of combs glued teeth up with wires attached to contacts on each side. The phone company installs its lines on one side of the block, and your lines are connected to the other side. The distribution frame will be located in some out-of-the-way place. Often, this location is called a wiring closet.

Network interface

All new telephone wiring done since January 1987 will have a network interface, a modular telephone jack used to help you or the telephone company determine whether the source of a problem exists in your inside wiring—or in the outside wiring somewhere in telephone company land. There will be an interface jack for each installed line. It is worthwhile for you to locate this equipment because you can use it to easily troubleshoot your telephone system and save the expense of an unnecessary service call.

Perhaps the easiest way to think about a network interface is to think of it like one of the circuit breakers in your main electrical panel. If you shut off a circuit breaker, electricity to the circuit is shut off. Likewise, if you remove the modular phone plug from a network interface jack, telephone service is shut off on that line. To turn electrical power back on, you just switch the circuit breaker to the on-position. To restore phone service to a line, you simply insert the modular plug back into the network interface jack.

Unfortunately, not all network interface devices look alike. Some are round plastic boxes and others are oblong boxes. Still others look exactly like the modular phone plugs you probably have along your walls or baseboards. And, if you have a small office, with only one modular jack, it's possible that the jack you are using *is* the network interface jack.

Types of Network Interfaces

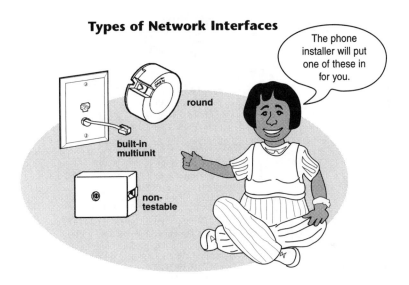

If you have more than one line, the network interface jacks will reside in an access box, often shaped like a cigar-box and made of gray plastic with a hinged cover. The phone company's lines, some testing circuitry and a ground wire are located in one compartment. This compartment is sealed by the phone company after installation. The portion of the box that you can access contains your telephone wires which are attached to terminals. It also contains your interface jacks. The wires will be attached to the interface jack with a very short run of telephone cord attached to a modular plug.

The access box is often just secured with a screw. To avoid worries about vandalism, wire-tapping, or pirated long-distance calls billed to you, it's a good idea to install a lock on this box.

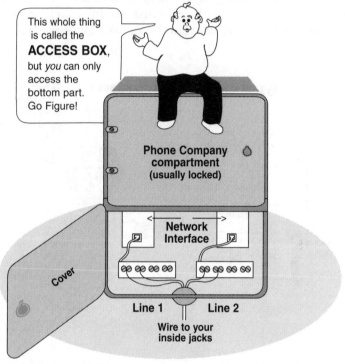

2-Line Network Interface
(Outdoor Type)

This whole thing is called the **ACCESS BOX**, but *you* can only access the bottom part. Go Figure!

Phone Company compartment (usually locked)

Network Interface

Cover

Line 1   Line 2

Wire to your inside jacks

Okay, now that you've found the network interface, what can you do with it?

• **Troubleshoot**
Use it to troubleshoot your system. If you experience trouble with your phone service, disconnect your service by removing the modular phone plug from a network interface jack. Plug a phone—that you know works—into the network interface jack. If the phone still operates, the problem most likely resides somewhere in your system. If the phone does not work, the problem is most likely caused by something on the telephone company side and you should contact telephone repair.

- **Prevent shocks**

  Use it to avoid shocks while working on your lines. If you work on your own telephone wires, you can unplug your network interface and avoid all possibility of electric shock. This is similar to turning off the circuit breaker on an electrical circuit before beginning to work on the wire.

Wiring plans

From the access box, the lines may run inside the walls, along baseboards, under the carpet, in the attic, or through the basement to the individual wall jacks where you can plug in equipment.

Most homes or small offices have one of two wiring plans:

- Series wiring
- Star wiring

Series wiring

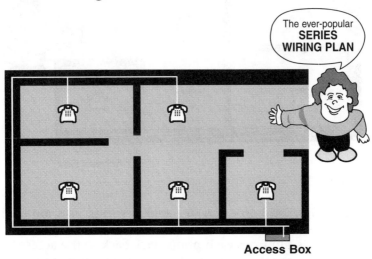

The ever-popular **SERIES WIRING PLAN**

Access Box

The majority of homes and about half of America's small offices have series wiring. In series wiring, the wires run from phone jack to phone jack. If you add a jack to the series, it is wired in line (often at the end of the line).

Although this method saves wire and installation time, it has a major drawback—no flexibility. Since all jacks are wired the same, if you want to add a line, you must rewire everywhere. Also, if a single jack fails, the entire phone system may fail. If, for example, you have five phone jacks, and the second jack in the series failed, all phone service from the second jack on would be out. This kind of wiring is somewhat akin to those cheap Christmas tree light strings that force you to remove each tiny bulb in succession to find the faulty one. And, until you do, the entire string is dead. Want to do that with your phone service?

Star wiring

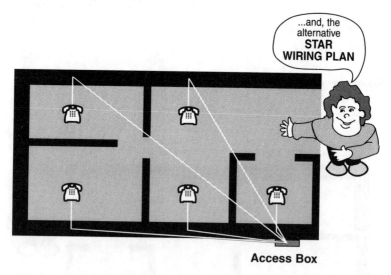

Access Box

Star (also called home run or parallel) wiring uses more wire but is the preferable solution. In star wiring, a separate wire (or wires) is run from each phone jack back to the access box. This allows you the flexibility to route your phone service in a variety of ways. For example, you might have two lines to your office phones, a separate line to the reception area, and a dedicated line to your copier room for your fax machine. Another advantage of star wiring is that if one jack fails, the remainder of your phones can still operate.

You'll need star wiring if you have many lines or wish to install a key system or a PBX. If you have a choice, go for star wiring. If your building is already wired serially, you can improve the chance of avoiding problems by doing all your new wiring as star wiring.

Phone wires

The majority of household and small-office wiring is "twisted pair." If you cut a cross section of the telephone wire you will find not two but four copper wires, each wrapped in a protective plastic sheath. Each wire is wrapped in a different color (usually red, green, yellow and black). This standardized color-coding makes it easy to rewire or troubleshoot telephone lines.

A telephone needs only two wires to operate. If there is only one phone line, the wires that are used are typically the red and green set. Yet almost every set of wires in American office buildings and homes has four wires. What are the other two there for? In the past, they were used to supply electrical power to phones that had lighted dials or intercoms. Nowadays, they are extra capacity. This is very fortunate because it means that you can install a second line into your home or office with no new wiring! Just connect the yellow and black wires, order a new line from your phone company, plug in a two-line phone and voila! You're in business!

Types of telephone wire

Extension wire

This is the line used to connect a modular telephone device to the wall outlet. This wire is often called satin wire. It is flat and often a silver color. If you cut this wire you will see that the color-coded wires inside are aligned side by side. You can purchase this wire in reels to make your own custom connections, or you can buy ready-made extensions of various lengths.

Cable wire

This is the line used to run telephone service throughout the inside of your building. This wire is round, thicker than satin wire, and usually gray. Inside, you'll find the color-coded wires. This wire comes in reels. If you are going to hire a professional installer to do your wiring, she will use one of two grades of wiring:

- **Level 3 cable** - standard voice quality telephone wiring. Use for voice communications and/or ethernet.

- **Level 5 cable** - higher quality telephone wiring used to support digital circuits up to 100 megahertz. Use for networking or high speed data needs.

Number of wires (pair)

Homes and small to medium offices today normally use one of the following:

- **2-pair wire**. This is the most common type of wiring found in homes today. You can install up to two lines using 2-pair wire. If you stripped the insulation off this wire, you'd find four wires inside.

- **3-pair wire**. This adds two more wires to the cable and increases capacity to three lines. If you stripped the insulation off this wire, you'd find six wires inside.

- **4-pair wire**. This increases capacity to four lines. If you stripped the insulation off this wire, you'd find eight wires inside.

Use the right wire for the right job

Some people might be tempted to use flat satin extension wire to replace round phone cable for interior wiring. Since it is slightly less expensive and works about the same, is it okay to use? No! Don't fall into this trap. Phone cable costs more for a reason—it's heavier-duty to stand up to years of use.

Mice and rats love to nibble on insulated wire. The thicker the insulation, the better chance you have to avoid having a connection severed by a mouse attack. And, more importantly, phone cable has twisted wires. This avoids the possibility of line interference and crosstalk (a faint conversation heard in the background). Though you can use flat, untwisted wire for short runs between your wall plug and your telephone connection, these are not suitable for longer runs.

Unfortunately, since industry deregulation, the use of non-twisted wire has become increasingly common, especially in prewired new construction. The results can have serious consequences for your business: garbled fax and modem transmissions, security leaks through overheard conversations, disjointed communication caused by line pops and snaps.

Leaky wires = lost business
A decorator with two lines lost a prospective customer when her caller, who was on hold, overheard her call to a vendor to get a wholesale price. When she relinked with her prospect and quoted a considerably marked-up price, the caller was livid and told her off in no uncertain terms. The culprit in all this: crosstalk leaking through the cheap wiring.

Fortunately for us, phone system designers made their wiring systems downwardly compatible. That means you can use higher capacity wire with equipment that needs fewer pairs of conductors and not lose service. You could, for example, operate a two-line phone using 3-pair wire. Or you could have 4-pair wiring and just use one-line service.

Save money: install extra capacity
If you're rewiring your office (or wiring it for the first time), wire with high capacity wire. Though you may only need two lines for now, it will only cost you a few dollars more to wire with 4-pair wire instead of 2-pair wire.

Always plan for more than you need
Example: You had your office wired for two lines using 2-pair wiring and you installed four phone jacks (one for your voice line, one for your answering machine, one for your fax machine, and one for the back of the shop). In the San Francisco Bay Area, it costs about $175 to add another cable run to your office and install a new phone jack. The bulk of the cost is labor. Materials are quite cheap. If you wired for two lines and four jacks, it would cost about $700. If you wired for four lines and four jacks, the only extra expense would be the cost of higher capacity cable and connectors (around $20).

Two years later you find that you need to add more lines because business is great and you hired three new associates. With some clever planning, maybe you could still get by with the same number of phone jacks (using splitters and switches, but your line capacity is too small. So, you have to replace all the interior cable with 3- or 4-pair wire. By "saving" $20 two years ago, you're out another $700.

Four-pair wiring gives you great flexibility. You could, for example, use two lines now for voice and reserve two for future computer networking (ethernet or token ring networks can run on phone lines and require 2-pair wire. Or you could have two voice lines, one dedicated line for fax and modem use, and keep the last as a spare for growth. Then, when you are ready to grow, you just call your phone company, they hook up your service and give you a new number or two and you're in business. It's that simple.

Types of connections

You'll most likely have one of the following types of phone connectors:

* Hard-wired
* Modular
* Four-prong

Hard-wired connection

This type of connection permanently wires a phone instrument to a wall plug called a connecting block. To connect another device, you have to cut the wire and either hardwire the new device or replace the connecting block with a modular jack. If you have any of this kind of equipment, learn how to replace the connections with modular connections or have someone do it for you.

You can purchase modular jack kits that replace older permanently-wired jacks for about $3. Your local hardware store, electronics store, or telephone specialty shop will have all the equipment you need to perform this task. The flexibility you gain is well worth the effort and minimal cost, especially if you do it yourself.

Modular connection

Modular Plug and Jack(s)

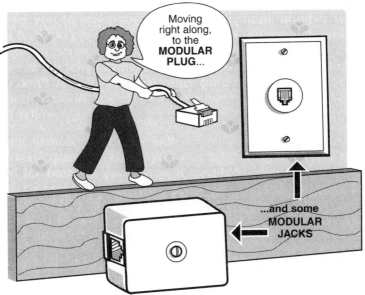

Modern telephone equipment uses a snap-together system of plugs and jacks to connect and disconnect devices with the telephone wiring. These connectors make it extremely easy to install or move equipment. All you do is press down on the plastic locking lever and insert the plug into the jack far enough that you hear it click in place. That's it—you're connected. To disconnect a device, simply press down on the plastic locking lever and gently pull the plug out of the jack.

Four-prong connection

Four prong (or 4-pin) connectors were the first interchangeable telephone connectors. Each plug consists of four long pins that are set in a particular pattern that you align with a four-prong jack to plug the phone in correctly. This type of connector is quite bulky, and it was soon replaced by the tiny plastic modular connectors you find today on modems, phones, computers, answering machines and the like. You can purchase inexpensive adapters that convert the older four-prong connection to a modular connection.

Not all modular connectors are equal
Modular plugs and jacks look alike and are interchangeable. However, there are slight but important differences among them. These differences control how many lines you can connect through one cord.

Plugs
You can determine how many lines your connector supports by looking through the plastic plug and counting the number of colored wires you see and dividing by two. Two wires = 1 line; four wires = 2 lines; six wires = 3 lines.

For our purposes, we'll pay attention to three types:

- **1-line.** Used for standard single-line phones, modems, answering machines and other accessories. This plug has two colored wires. The phone company calls this connector an RJ-11 plug.

- **2-line.** Supplies phone service for one or two lines. It looks exactly like the 1-line except that you see four colored wires inside the plug. The phone company calls this connector an RJ-14 plug.

- **3-line.** Used to supply phone service for one to three lines. It looks similar to the 1 & 2-line plugs, except that the cord is wider and it has six colored wires. The phone company calls this connector an RJ-25 plug.

Jacks
If you peer inside a jack, you'll see anywhere from two to eight gold stripes (called pins or contacts). These contacts indicate the potential number of lines this jack can accommodate. To determine how many lines are actually connected at this jack, you'll either have to look at a wiring blueprint of your office or take the cover off the jack and count how many wires are connected.

Jacks and plugs must match
The contacts inside the plug must align exactly with the contacts inside the jack. Otherwise, not all your phone service can get through. For example, if you have two phone lines and you use a 1-line plug to attach a two-line telephone to

your wall plug, only line one will work. That is because the 1-line plug has only two contacts and is wired for one-line service. While another plug may appear to fit into the outlet, unless it has the proper number of contacts aligned the way your system needs them, your service either won't work or won't work to capacity.

Maximizing your current setup

Splitter
Want to add a phone accessory like a Caller ID Display or an answering machine at the same location as your voice phone, but you don't have a spare phone jack available? Just use a splitter to convert a single outlet to a multiple outlet. Plug

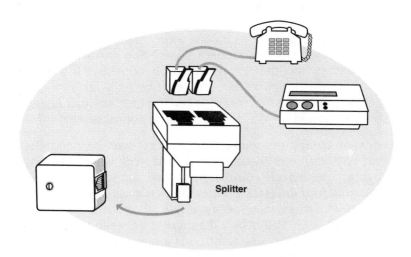

Splitter

this inexpensive device into your phone outlet, and plug two cords into it—one for your phone, the other for your accessory. Splitters come in two, three and even five-plug combinations and start at around $3. These devices come by many names: splitter, outlet coupler, two-and three-way jack, outlet modular adapter, etc. You get the idea.

Wireless phone jack

Need a jack somewhere else but can't or don't want to do a permanent installation? With a wireless phone jack system, you can plug a jack into any electrical outlet. The system comes with a base unit that you plug into a working phone jack, and an extension jack that you plug into an electrical outlet. Then you just plug your phone into the extension jack. When you want to move the jack, just unplug it. You can get these at specialty phone stores. They're not cheap (about $100 for a base unit and one extension jack, $50 for each additional extension), but they sure are easy. *Note*: These may not work well with fax machines or modems.

Line sharing

Have only one phone line to share between your fax machine, answering machine and your telephone? Want your line to be able to recognize what kind of call is coming in and route it to the appropriate device—to your phone (or answering machine if turned on) if it's a voice call, and to your fax if it hears those screechy modem tones? You can purchase an automatic switch that routes the calls correctly. If a call is made to your line, you can answer it or let your answering machine get it. If a fax signal is on the line, the call is automatically routed to your fax machine or fax/modem.

Loads of companies make line sharing switches and they come with different features. Some can work with Distinctive Ringing, a telephone company-supplied service that lets you have up to four different phone numbers ring on one line (see Chapter 3). Others can specify which phone number should ring at each jack; still others distinguish between modem and fax calls. An inexpensive solution is a manual switch that lets your turn a knob and switch between two devices on a single line. Numerous companies make line switches. Check out these handy catalogs for a range of line switches:

- Hello Direct
  800-444-3556

- Radio Shack
  800-843-7422

Troubleshooting

Troubleshooting a telephone problem is usually done by the process of elimination. Ask yourself these questions:

1. When did this begin to happen?
2. Have I installed any new equipment?
3. Can I receive calls okay but can't dial out?
4. Can I dial out okay but can't receive calls?
5. Has anyone been working elsewhere inside/outside the building? Phone lines can easily be accidentally cut by electricians, plumbers, street workers, etc.
6. Could weather be affecting service?
7. Check the simplest things first:
    ✓ Is the device plugged in?
    ✓ Is the ringer turned down or off?

Pay attention to safety

Telephones are electrical devices. Phones always have low-voltage electrical current on them—enough to give you a mild shock. High-voltage current is generated when the phone rings. If you're touching live wire when the device rings, you will get a sizable shock (105 volts). To avoid these hazards, follow these simple safety rules when working on telephone equipment:

- If you wear a pacemaker, you should avoid all contact with live telephone wires.
- Do not work on active telephone lines while standing in water.
- Do not work where there is a risk of dropping a telephone instrument or wire into water.
- Avoid working during thunderstorms.
- Use shielded tools.
- Unplug your service at the network interface jack or take a telephone off the hook before working on telephone wires. This will "busy-out" your phone line and you'll avoid the possibility of a dangerous shock if a phone should ring.

A few common problems

Other phones or devices using this telephone line work, but this phone doesn't work

Take the suspect phone and try it out in a working jack. If it works, great! You can assume that the jack at the previous location is bad. If the phone still doesn't work, you now know that there is something wrong either with the phone itself or with the line cord.

Nothing works on your line

If you know where your network interface is, unplug the network interface jack. Then plug in a telephone that you know is working into the network interface jack. If you hear dial tone and can dial out, the problem is on your side. If the line is dead, call phone company repair—the problem is on their side.

Shared phone line, one faulty piece of equipment

If you have more than one device on your telephone line (modem, fax, fax/modem, phone, etc.), an electrical short in the device could trip up your entire system. Unplug all phones and accessories on that line. Replace a device at a time and see if you can get dial tone. Continue until you experience failure—that device is the culprit.

Last year, I experienced a shorted-out line. At the time, I had four extension telephones on one voice line. A fax machine shared line two with my computer modem. One day, when I was expecting several calls, I noticed that line one wasn't ringing. When I picked up the line to check for messages, I was surprised to find that my phone company voice mail had several messages waiting for me, which was strange because I had never left the office all day and I had not been on the line. So I called my voice line from my fax line and, guess what? No rings on the line at home and, after four rings, voice mail picked up.

I called the phone company. It showed no problems reported at its end and suggested that some of my equipment might be shorting out the line. So I checked each phone in turn and found that whenever I plugged in a particular cheapo phone,

I'd lose all ability to receive calls. Once I unplugged that phone, service seemed fine. Needless to say, I banished that phone.

Of course, you still don't know if the problem resides with the line cord or with the phone or the accessory itself. But, if you have a spare working line cord you can test this out too. Or, you can take the phone, accessory and line cord to your local telephone accessory store. Many of them offer free phone and line cord testing service. It only takes a few minutes.

No ring or soft ring

If you don't hear your phone ring, or the ring is very soft, and you've checked that the ringer is turned on, you may have too many devices on the line. Your local phone company guarantees to supply only enough power on the phone line to ring the equivalent of five standard telephones. If you have a lot of extensions or other equipment that rings (like answering or fax machines) on the same phone line, you might max out the ringer electronics. To fix this situation, simply turn off the ringer on one or more devices and see if ringing is restored.

Intermittent problem—line two sometimes doesn't work

This is a rather unusual problem, but it can happen if you have two lines and are using a Princess or Trimline phone. The symptoms: When you pick up the telephone, you can make calls normally. But while the Princess or Trimline phone is off the hook, line two acts dead. You can't dial out on it, you can't receive calls, and if someone calls you on the second line, they get a busy signal.

The problem is that Princess and Trimline phones use the wires for your second phone line to light up the phone dial. When you use the phone, all four wires are used. What can you do about it? Replace the old style phone, use a 2-pair instead of a 4-pair line cord to attach the phone to the wall jack, or convert the phone's wiring to standard 2-pair.

Do-it-yourself wiring

Running telephone wiring is quite simple. In most systems, you'll find four wires (called a twisted pair) coming to each phone jack. Only two wires are needed for each telephone line. Phone wires are color-coded so it's hard to make a mistake. Be sure, however, to follow safety instructions and use the correct wires, plugs and adapters for the job.

If you want to do it yourself, you'll find loads of help at your library. I'll mention a few useful resources on the next couple of pages.

Resources

The Phone Book by Carl Oppendahl and the Editors of Consumer Reports Books
Consumer Reports Books, 1991
101 Truman Ave.
Yonkers, NY 10703
This fact-filled book covers long-distance and local service, in-home repair, installation and wiring, and choosing equipment.

All Thumbs Guide to Telephones and Answering Machines
by Gene B. Williams
Tab Books, a Division of McGraw Hill, Inc., 1994
Blue Ridge Summit, PA 17294-0850
Full of illustrations and refreshingly brief, this guide covers safety, basic tools needed, and basic trouble-shooting for telephone wiring, telephones, cordless phones and answering machines.

Do Your Own Wiring
by K. E. Armpriester
Popular Science
Sterling Publishing Company
387 Park Ave. S.
New York, NY 10016
Contains an excellent chapter on telephone wiring in an easy-to-follow visual format.

Installing Telephones
by Gerald Luecke & James B. Allen
Radio Shack

Master Publishing, Inc., 1992
14 Canyon Creek Village MS 31
Richardson, TX 75080
This useful booklet contains clear and well-illustrated instructions for making quick modular telephone replacements, adding or changing telephones, running cables, and installing single and multi-line telephones.

Installing Your Telephone (video)
Radio Shack, 1994
same address as above
This half-hour video covers the following topics: modular connections, converting older wiring systems, adding a new outlet, installing round jacks, installing jacks near electrical outlets, running telephone wire, installing wall telephones and troubleshooting.

3...Phone Services

• •

During the last decade, the number and variety of phone services has increased dramatically. You can now use phone services to discover who's calling (Caller ID), provide several phone numbers that all ring on the same line (Distinctive Ring), program your phone to find you anywhere (500 service), or earn you money every time someone calls your pay-per-call number (900 service). You can call back the person who just called and hung up, even though you don't have a clue who that was (Last Call Return) or set up a special ring pattern that lets you know that your best customer is calling (Priority Ringing).

Perhaps the last time you paid any attention to phone company services was when you initially ordered phone lines for your home, office or business. You may have signed up for hunting or call waiting—maybe even voice mail. Unless you knew what to ask for, it's unlikely that your sales rep discussed most of the services available. The purpose of this chapter is to explain these services and give you some idea how you might use them in your business.

How phone services work

Telephone services are mostly computer-driven. Because your telephone calls go through a computer switch in the telephone company's central office, it's relatively simple to add services (such as voice mail or Caller ID) to your line. The call-handling services that are supplied to that line are stored on the central office computer.

Most of these central office switches are fully digital, meaning that telephone conversations are converted to digital form before being passed through the switch. They're later reconverted back to analog and sent to you. Digital technology allows local phone companies to process calls faster, offer equal access to competing long-distance carriers and store more information about each phone line (such as the call handling features supplied to that line). The additions, deletions and changes to the service on the line are done by changing a few characters in a line of software code. Because features now are often software-controlled, they are much easier to set up and deliver.

Some phone services require additional equipment to take advantage of a feature (such as Caller ID), but most are designed for use with a standard phone using the touch-tone keypad. Many of the features can be turned on or off by dialing a feature code—usually an asterisk followed by a one- or two-digit number. Rotary phone users prefix their codes with an 11 instead of the asterisk. Many of the codes require a *hookflash*. Depending on the type of telephone set you have, you perform a hookflash by pressing and releasing the receiver button or switch hook that sits under the handset or by pressing a flash button.

The codes that work these special features are not the same throughout the United States. For example, in most parts of the country, pressing hookflash puts your caller on hold and lets you answer a call waiting. However, in New England Telephone territory, you must dial * 9 to do the same thing.

You pay for phone services monthly. Often, if you buy more than one service or feature, you'll receive a discounted monthly service fee. If you order the services when you initially order a new phone line, the installation fee may be waived. If you wish to remove the service later, there is usually no disconnect fee. However, if you wish to add a service later, there will most likely be a fee.

Of course, these features work on your land lines only. If you have cellular service, you'll need to check with your wireless vendor to see what is available over the air.

My favorite phone services

There are loads of phone services to choose from. First, my favorites:

Distinctive ring

Assigns multiple phone numbers to the same line. Each phone number rings with a different cadence so you can tell what type of call is ringing in—a business call responding to your radio ad, an urgent personal call (you give out a special number for emergencies) or a fax transmission, for example. You can then print your fax number, voice number and an emergency number right on your business card. Only you know that they all ring on the same line.

This service goes by many names throughout the country including RingMate, Identa Ring, Ring Master, Personalized Ring, Custom Ringing, Route-a-Call, Teen Service, Multi-Line and Smart Ring. You pay a monthly fee per number (ranges from $4 to $7 per number), a lot less than a separate line.

If you want to use this service and hang various devices on the line (e.g., phone, answering machine, fax modem, etc.), you'll need a switch with distinctive ring capability. The switch plugs into the phone line and has several modular jacks into which you plug your devices. When a call comes in, the switch "listens" to the ring pattern and connects the call to the correct device. Some answering machines and voice mail devices also support distinctive ringing. You can find these switches at any telephone or electronics store or through a telephone supply catalog.

Incidentally, if you have Call Waiting, it will also ring a corresponding distinctive tone (e.g., two, three or four beeps) when a call comes in for the associated number.

A note of caution: If you plan to order this service, be very specific with your phone company and ask them to explain how it works. Because this service has such a variety of names, you may find that your local phone service offers something else but calls it distinctive ringing. Some telephone companies sell a distinctive ringing service that rings a special cadence, but it is just for ten special callers on a list you program into your phone.

Caller ID services

Caller ID lets you see the phone number (and, for an extra fee, the caller's name) of the incoming call before you answer the phone. Caller ID is usually a combination of two features: Calling Number Delivery and Calling Name Delivery. Calling Number Delivery displays the 10-digit telephone number of an incoming call. Calling Name Delivery displays the name (as it is listed in the directory) associated with an incoming call. Both services display the date and time—a convenience especially if you are automatically logging calls. The service costs range between $4 and $7.50 per month, depending on the service provider.

Caller ID is useful for call screening—especially if you link it to an answering machine or voice mail. Then you can minimize interruptions, monitor calls discreetly, and decide which calls to take now, and which to return later. Software vendors are developing Caller ID applications that do "screen pops," by selecting a customer's record from the database as the call comes in and displaying a screen full of data such as order history, terms, etc. This allows you to be fully prepared to handle the caller's needs.

Several software vendors sell Caller ID products. Some work with contact management software; others let you set up a separate database. Some possibilities:

- **Call Editor**
  VIVE Synergies (905-882-6107). Works with contact-management software such as Symantec's Act!

- **Sidekick (Windows and Mac versions)**
  Starfish Software (800-765-7839). This well-known personal organizer program now comes with Caller-ID capabilities. Requires a Caller-ID interface device.

- **The Bridge to Caller-ID**
  Postek (800-767-8351). This works with contact managers such as Act! and GoldMine.

You can create a database of all your calls—how long they take, where they call from, what they ordered. Even unanswered calls are captured, so you could call them back even though they didn't leave a message.

False delivery orders are stopped cold by Caller ID. Delicatessens, pizza parlors, taxicab companies, and the like can verify that the phone number the caller provides actually matches the number displayed. Domino's Pizza reports that Caller ID has helped it reduce the amount of bad orders by more than 90%.

Fewer robberies
Mickey Schiavone, manager of Victory Taxi in New Brunswick, said that cab robberies fell more than 80 percent since his company started using Caller ID. "It also helps in understanding people who speak another language when they give an address," Schiavone explained. (*Asbury Park Press*, New Jersey Take-Out Services Rely on Caller-ID to Screen Prank Orders, 1/26/95)

To take advantage of Caller ID, you need to sign up for the service and install a Caller ID translator. This could be a phone with Caller ID display capabilities, a separate Caller ID device or a telecommunications board installed in your computer with Caller ID software. Caller ID-equipped phones can be found in prices ranging from $75 to $175. If you don't need a new phone or don't want to pay those prices, you can order an external Caller ID display unit from Northern Telecom (800-667-8437) or Hello Direct (800-444-3556).

Caller ID information will display only if the information is available. If the call comes from an area that does not support Caller ID, the display will show "OUT OF AREA." If the caller has blocked Caller ID by using a privacy code (such as *67), you'll see "PRIVATE." If the call is coming from someone with a PBX, the number displayed will be the customer's main billing number, not the actual telephone number of the person calling.

Note: Caller ID information is delivered between the first and second ring. That means that if you pick up too quickly, you won't receive it. Wait until after the second ring to answer.

The issue of Caller ID has long been enveloped in controversy. Oddly enough, both sides cite privacy as one of their principal arguments. Opponents maintain that Caller ID takes away their right to make an anonymous call or to remain unlisted and gives telemarketers and other computer databases too much information about them. Proponents include people who want to use Caller ID as a call-screening device, as well as businesses who see the advantage of knowing who's calling. To help resolve the issue, the FCC has mandated that Caller ID Blocking can be enabled before any phone call by pressing *67 before dialing out.

Forwarding calls

With a call forwarding feature, you can forward—or transfer—your incoming calls to any number you choose (even your cellular phone). When calls are forwarded, all calls ring at the forwarded-to number instead of your regular number. Use it to forward your calls from the office home or vice versa. Send calls to a second line. Forward your fax calls to a fax modem when you're on the road.

When Call Forwarding is activated, all your calls are forwarded immediately to the number programmed. If a call comes in for you, your phone may ring about a half a ring (called a reminder ring), then forward to the number specified. Callers won't hear a busy signal when you are on the line. However, if the number to which you are forwarding is busy, your callers will receive a busy tone at the forwarded number. You can still make outgoing calls when Call Forwarding is activated.

Call Forwarding comes in three basic types:

- **User-programmed forwarding**.
  Gives you the ability to program your number to automatically transfer all incoming calls to another number.

- **Busy forwarding**
  Calls forward when your line is busy.

- **Delay forwarding**
  Calls forward when you don't answer after a predesignated number of rings. This gives you a chance to answer the telephone first.

Note: In many areas you cannot program Call Forwarding remotely. Other areas offer a service that allows you to forward calls remotely by dialing an access code and personal identification number (PIN).

If you're absent-minded, Call Forwarding can be inconvenient. If you forget that you've forwarded your calls, your calls will continue to ring at the wrong site until someone notices and cancels the call forwarding instruction. Some smart phone systems have a reminder line or message indicating that calls have been forwarded.

In addition to the monthly service charge, you pay normal usage charges when calls are forwarded. The charges are based on the distance from your central office to the number to which you are forwarding.

Toll-free service

Provides your callers with a toll-free number which encourages incoming business and provides a professional image. According to Jeffrey Kagan of the *Atlanta Business Chronicle* ("Making the Right Call," September 1994), 86% of customers believe that an 800 number connotes high quality products or services. "No small business should be without an 800 number," he advises. If you see several ads for the same product or service, don't you call the 800 numbers first?

The number can be customized to receive calls from a particular area (by prefix, area code, state or, in some cases, nation). It's easy to adjust the size of this area whenever you want—which is useful for target marketing and seasonal campaigns. The 800 (or 888) number can be programmed to ring in on (or "piggyback" onto) your current number. No on-site installation is required. If you expect a high volume of calls, you can set up a separate line just for toll-free calls.

Some entrepreneurs provide their toll-free number to a select group of customers for premium service. Others print it on their business stationery and display it on every ad.

You can program your toll-free number to route calls to a different location. For example, you could send your calls to your business location during office hours and to voice mail (or your home) after hours. You could even have different routing schemes for different days of the week. When a caller calls an 800 or 888 number, the computer at the central office switch looks up the actual number that is associated with the toll-free number and forwards the call.

Some organizations set up an emergency routing scheme for their toll-free number that they hope they'll never use. In the event of a natural disaster or other catastrophe that prevents employees from getting to work, you call your phone company and ask that the alternate routing plan be used. This could, for example, send all your calls to a voice announcement service, voice mail or an alternate location. The alternate scheme stays in memory until invoked.

You can combine toll-free service with distinctive ringing. You'll know, when the call rings in, whether you're paying for it or they are.

A side benefit of toll-free service is that you are listed in the toll-free database. Customers may locate you by calling 800-555-1212. Many long-distance carriers publish toll-free directories in several different versions (international, Spanish language, consumer, business to business, etc.). If your number is for private or restricted use, you can ask that your 800 or 888 number be unlisted. During a brief scan of AT&T's Toll-Free Directory, I located a supplier of Amish quilts, a handwriting analysis service, a referral service for black-owned businesses, a 24-hour crab cake delivery hotline, a dial-a-lawyer advice line, a home-for-sale locator service, a network of therapists... you get the picture.

Toll-free service costs vary considerably. Some vendors charge a monthly fee; others do not. Usage charges based on call duration range from 7.5¢ to 29.9¢ a minute depending on the time of day. Rates vary considerably depending on the area of your 800 number reach. Many services come with other attractive options like voice mail, fax-on-demand, and follow-you-everywhere service.

In some cases, you can get the best deal from a reseller. A reseller is a long-distance company that does not have its own transmission lines. It buys lines from long-distance carriers at a discount and resells them to its customers. Be sure to shop around.

Follow-me phone service

Portable one-number phone service (sometimes called 500 service) that follows you everywhere is especially useful for road warriors. Your calls can be forwarded to any number that you can directly dial, including cellular phones, fax machines, pagers and voice mail.

> **On-the-go service**
> Ron Kopp, a management consultant, is always traveling. His 500 number service has the ability to program three numbers that are tried serially. If there is no answer at any of the numbers specified, his calls are forwarded to his home answering machine. While the phone service is tracking Ron down, his callers hear "please stand by, we are trying your party at another location."

One service, AT&T's True Connections, lets you set up a schedule for call forwarding. For example, you might forward calls to your cellular phone between 8:00 and 9:00; your office from 9:00 to 5:00, your home phone from 6:00 to 10:00, and to voice mail after 10 p.m. Billing is handled two ways: Calls forwarded to you are billed to you if the caller keys in a special PIN code (which you've provided); otherwise the caller pays for the forwarding charges. You can sign up for this service only if you have AT&T long-distance service.

Some carriers are offering follow-me services as an 800 service option. For example, MCI offers a service that links all your messaging services—standard telephone, cellular service, fax, voice mail and paging—to a single 800 number. You can program a desired routing plan for incoming calls with up to three different numbers that will be tried in order. At each stage in the sequence, callers are advised through an automated voice prompt that the system is still trying to locate you. If your call is a fax, the system recognizes the fax tones and directs the call to a fax mailbox for storage and later retrieval.

You can also arrange with your local telephone company or cellular vendor for follow-me service. It's marketed under a variety of names including ContactLine, ProLink, New Vector, Total Number, MyLine and Personal Number.

900 (and 976) service

On the final night of the Reagan-Carter presidential debates, ABC-TV offered a phone poll service through which callers could call in and vote their preference. A half a million people paid 50¢ each to voice their opinion. Although your 900 idea probably won't generate that kind of response, there are opportunities to do very well in this industry.

You set up a special telephone line that dispenses something people want (like information or entertainment) and charge them to use it. 900 numbers are national; 976 numbers are regional. Although 900 service got badly tarnished by the dial-a-porn programs of the '80s, it's a viable business that offers hundreds of legitimate and useful information services such as:

- Software support help line
- Pet poison control center
- Legal advice
- Worldwide weather
- Pharmacist advice
- World trade opportunities

And then there are entertainment services like:

- Music concert connection
- Soap opera summaries
- Michael Bolton's fan club

According to Robert Mastin, author of *900 Know-How*, you can start a 900 business from home with a fairly small initial investment (often around $2,000). By using a service bureau, you don't have to purchase any special equipment. Operating costs vary depending on the call volume generated. Transport fees range from 28¢ to 32¢ per minute. Fees for regional 976 numbers are lower—10¢ or less per minute. And, if your program is successful, you can be earning revenues 24-hours-a-day, seven days-a-week, any time someone calls.

The trick is getting them to call. You must offer a unique service at a reasonable price, and get the word out. That means lots of advertising and promotion. The turnover in 900 service is high. According to Mastin, only 5% to 10% of new 900 programs succeed after the first six to twelve months.

Note: To make it in the 900 business, you need to do your homework. For a list of sources to help you, see *Resources* at the end of this chapter.

A phone service sampler
You may not find all the services and features listed here offered by your local phone company. Some of the more advanced features require digital switches or expensive software upgrades and may not be implemented throughout your region. Those of you residing in urban areas will probably get the newer stuff first.

Monthly costs for these services range from $2.50 to $7 per service. Check with your local phone company for the exact price in your area.

Anonymous Call Rejection
This service works in conjunction with Caller ID and allows you to automatically reject all calls from callers who have activated blocking Caller ID. Your caller receives a recorded message advising that blocked calls are not accepted. Some residential subscribers like this service. Not too handy for the normal business.

Block Call
Depending on your area, this (1) lets you prevent certain types of outgoing calls, such as toll calls (you can override blocking by dialing a personal code) or (2) allows you to

program a list of numbers that will be blocked from ringing on your phone line. See *Call Screening.*

Call on My Dime

You arrange for a personal four-digit number that you give
to preferred customers and business associates. Whenever
or wherever they call, the charge is billed to you. Bell Canada's Call-Me service works like this: callers dial zero and
simply key in the four digit code. Easy!

Call Pickup

Lets you answer a call to one line by dialing the call
pickup access code from another phone and answering
the call at that location. This feature is useful if you have a
plain telephone set with no special buttons and want to
answer calls for coworkers without having to get up from
your desk and walk over to their phone to pick it up.

Call Screening

You program a list of phone numbers that you absolutely
never want to talk to again. This is not very useful for
most businesses; but some residential customers, home-
based businesses and non-profits use this service to get rid
of crank callers. Screened callers hear a polite message telling them you're not taking their calls. Period. You can
also add the number that just called you. You don't have
to know the number. Just press a special code.

Call Waiting

This feature informs you when someone is trying to reach
you while you are on the phone. While on the phone, a
beep tone alerts you to another call. If you press hookflash
quickly (or, in some parts of the country, dial a short
code), you put the first call on hold and can speak to the
incoming caller. To return to the previous caller, you just
hookflash or press the code again.

If you have Caller ID service, you can link it with Call
Waiting and see at a glance who's trying to reach you. You
decide whether you want to take the call.

Call Waiting has its fans and its foes. Though it gives you the capability to handle two calls at once, many people dislike being placed on hold and are especially unwilling to remain holding during a business call. Others think Call Waiting appears tacky and low-budget. This may not be the image you want to portray.

Call Waiting signals disrupt fax and modem transmissions, often causing your modem to hang up, a costly problem. If you have Call Waiting, you should get a separate line for data equipment or, at the very least, turn off Call Waiting before you initiate a fax or modem call.

Call Waiting also has problems with the Busy Call Forwarding feature used to send calls to voice mail when you're on the phone. When your line is busy and you have phone company voice mail and Call Waiting, none of your overflow calls will reach voice mail. See Chapter 8 for a complete discussion of this problem and some possible solutions.

You can cancel Call Waiting before making an outgoing call. This is usually done by pressing asterisk (*) 70 before dialing the number (1170 for rotary phone users). Note: This only works for the next outgoing call. It also does not protect incoming calls (such as faxes) from being disrupted.

Centrex

This is a central-office based phone system where you purchase business communication features like Call Transfer and night service as you need them. Available for businesses with two lines or more. See Chapter 7 for a detailed description.

Complete the Call

This works in conjunction with your phone company's Directory Assistance service. Callers to Directory Assistance hear a message offering to dial the number for them, free of charge as a courtesy, or for a stated fee.

Distinctive Ring Selective Call Forwarding

Works with the Distinctive Ring family and forwards only the calls dialed. You could, for example, forward faxes to a fax mailbox, business calls to your office answering machine, and personal calls to your mother-in-law's line.

Do Not Disturb

Lets you get work done without being constantly interrupted by a ringing phone. You can either turn this on and off at will (by pressing *78) or set up a regular schedule in advance. No calls before noon sounds great to me! You can give family members or preferred clients a privileged "caller code" that lets their calls ring through anyway.

Emergency Service

If you lose phone service for any reason, this will automatically reroute your calls to another number. Great for disaster recovery! If you rely on your phone service for your daily bread and butter, a service like this makes sense.

Extra Line Just When You Need It

This service lets you expand your phone service on the fly. For example, you may use this line only for faxes, or to get an outgoing line when all your lines are full. You pay only when you use it. US West's service is called Stand-By Line—it comes with a monthly summary so you can analyze your call traffic and forecast future needs.

Hold Call

Lets you place a call on hold even if you don't have a hold button on your phone. You do this by pressing hookflash, dialing the Call Hold code and hanging up. The call is placed on hold and is resumed when you pick up the handset at the original phone or at an extension.

Hunting

If you have more than one line, they can be programmed so that calls ring on another line if the first line is busy. The phone system "hunts" or "rolls over" to another line.

If all lines are busy, the caller hears a busy signal. This service is available only on business lines. You can't hunt between residential and business lines, even if they are in the same facility. To do that, you'll need to use call forwarding features.

If you design your hunting system correctly, it can be a very useful tool. However, if you add features to your line that rely on Busy Call Forwarding technology (such as voice mail), you may get undesirable results. For example, if you have two lines that hunt in sequence, and you put voice mail on the first line, your calls will always go to voice mail if the first line is busy and not roll over to the second line.

Both the Hunting feature and the Busy Call Forwarding feature instruct the computer switch how to handle your busy calls, and only one instruction can "win." Busy Call Forwarding takes precedence over hunting. If you want to combine Hunting with voice mail, be sure to talk to your telco rep, who can help you set up a system that should work for you. Prices vary greatly depending on your region. It's even free in some states.

Make Set Busy

Busies out your phone without having to take it off-hook. Especially useful for the telecommuter who needs to ignore incoming calls to complete a project. Of course, you'll need to have some alternate means of getting messages, such as voice mail.

Priority Ringing

You program in a list of phone numbers which will ring with a distinctive ring. In some areas, you can have only 10 numbers on this list. Other areas allow more numbers. This service is also called Priority Call and Call Selector.

Priority Ringing does not work if the person calling you is using a long-distance carrier. That is because the long-distance call does not carry the coded number in a manner that allows the local switch to read it. Priority Ringing does not work with business lines that are part of a PBX or Centrex system. That is because the number that is sent to the central office computer is the company's main number, not the number of the extension dialing out.

Repeat Dialing

Redials the last busy number you called and keeps trying until the line is free (for up to 30 minutes). If you dialed a number and reached a busy signal, you press a code. When the line is free, your phone will announce the call with a distinctive ring. *Note:* will not work if you use a carrier (e.g., long-distance phone company) other than your local phone company to place the call.

This won't tie up your phone line like a redialer on your phone would. You can still make and receive calls. When the system detects that the desired party's line is clear, your phone will ring you. Some services call this Continuous Redial.

Return Call

Automatically calls back the last number that called you, even though you don't know the number. If the line you're dialing is free, your call will be completed. If it is busy, this service will check the line for you for up to 30 minutes and let you know when the line is free by ringing you with a distinctive ring. Nice time saver. Missed calls can easily be returned. This function is also called Call Cue, Call Return and Last Call Return.

Note: Return Call does not work in all situations. It works if the person calling you was calling locally and was not using a phone system such as a PBX or Centrex system.

Select Call Forwarding

Allows you to forward a list of calls to an alternate number, and to restore the line to normal operation at your discretion. You program the list yourself and can update it whenever you choose. The calling party is not aware that the call is being forwarded. Select Call Forwarding can be used for screening calls. Also called Preferred Call Forwarding.

Selective Call Acceptance

Allows you to specify a list of up to 32 phone numbers from which calls will be accepted. When you activate this feature, all other callers hear a polite announcement telling them that you're not accepting their calls. By the way, the operator can ring through in case of an emergency.

Speed Dialing

You program a list of phone numbers that you can dial using a one- or two-digit code. Some phone companies allow six numbers only; others offer 8, 30 or 50 numbers. To use, simply press the code. Frankly, phone service supplied-speed calling is rarely cost-effective. If you want speedy dialing, get a phone with this feature or, better yet, an autodialer program on your PC.

Three-Way Calling

Allows you to link three phone lines (two plus yourself). Your secretary, for example, could use Three-Way Calling to send your calls to your home when you're working on a project. Your office would be billed for the forwarding charges to connect the caller to your home phone.

How it works: Dial the number of the first party you wish to link, tell them what you're up to, then press hookflash quickly. This puts the first party on hold and gives you a second dial tone. Dial the second number, tell them that you're going to connect the first party into the conversation, and press hookflash again. This puts you all together.

If you have Three-Way Calling service, you can also use it to place a caller on hold, and place another call while holding the first. This lets you use it as though you had an extra outgoing line.

Transfer Call

Gives you the ability to transfer a call from your phone to another phone. This saves you from having to ask the caller to hang up and dial the other number.

Internet calling

Several companies are offering voice telephone conversations over the Internet. You can make a phone call to virtually any-one anywhere, for as long as you wish, and all for the price of a local phone call.

What you need: A multimedia computer with a sound card, built-in microphone and speakers, a fast modem, subscription to the Internet and calling software. To make a call, you log on, select your phone software and click on the name of the person you want to talk with. That person must also be logged on at the same time, and using the same software to receive the call. The Internet phone service cannot call a conven-tional telephone at the other end. You can also make a call by typing an IP address to connect to a specific computer.

There are several software vendors that sell Internet phone software, Often you can download a test version for free from the Internet.

- Quarterdeck, WebTalk (800-683-6696)
  www.quarterdeck.com

- Vocaltec, Internet Phone (201-768-9400)
  www.vocaltec.com

- Third Planet Publishing, DigiPhone (800-442-7120)
  www.planeteers.com

- Microsoft, NetMeeting (800-426-9400)
  www.microsoft.com

If Internet calling sounds too good to be true, you're right. The interface is closer to CB radio than to a real phone. In most cases, the sound quality resembles a half-duplex speak-erphone. If two people speak at the same time, you cancel each other out. If you want better sound quality, look for soft-ware with full-duplex capability.

During high-traffic times on the Internet, you'll also have to contend with delays of up to one-second between the time you speak and the time you hear yourself speaking. Speech breaks up, syllables are dropped and sound is very choppy. Still, if you want to chat with someone across the world, and you're not fussy about sound quality, the price is right.

If you want to have a good Internet phone connection, have a high-speed computer, a very fast modem, a good sound-card and a headset. For top quality connections, sign up for an ISDN line (see Chapter 4).

In the final analysis, however, the Internet is not designed for voice traffic, and your best connection will be with a regular telephone over the telephone network.

On the horizon
Telephone services are proliferating. Among the most intriguing new services are:

Voice dialing
This lets you program your own personal telephone directory by saying a name in any language and keying in the corresponding phone number. Henceforth, you just pick up the phone, say the name ("broker" or "mom") and the call is automatically dialed. No special equipment is needed; this service works with rotary as well as touch-tone phones.

Spoken Caller ID
Look for Caller ID to speak soon. With this service, the caller's name is announced over your receiver, letting you know who is calling before you accept the call. You can then use voice commands to take the call, refuse it by playing a pre-recorded message or routing the call to your voice mail.

Resources

900 Know-How: How to Succeed With Your Own 900 Number Business. 3rd Edition
by Robert Mastin
Aegis, 1996
796 Aquidneck Avenue
Newport, RI 02842
800-828-6961

This is the authoritative book on setting up a 900 business. The author writes from experience—while researching his book, Mastin actually created and ran a profitable Dial-a-Joke 900 service.

Audiotex News
2362 Hempstead Turnpike, 2nd Floor
East Meadow, NY 11554
800-735-3398

Leading newsletter for the audiotext industry.

Money-Making 900 Numbers: How Entrepreneurs Use the Telephone to Sell Information
by Carol Morse Ginsburg and Robert Mastin
Aegis, 1995
800-828-6961

Contains profiles of nearly 400 different 900-number programs, including the winners and the losers.

Phone Company Services: Working Smarter with the Right Telecom Tools
by June Langhoff
Aegis, 1997
800-828-6961

Describes phone company services in detail, and how to put them to best use in real-life applications.

4...Phone Lines

It used to be so simple. All phone lines were the same. You just plugged in your phone and you were ready to make a call. Today, you have more choices. You can arrange for a line that will turn one little copper pair wire into two lines (ISDN) or 24 lines (T-1). You can set up a line that rings in two locations (off-premises extension) or give you a branch office without moving an inch (Remote Line). And, for a premium, you can set up super-fast connections.

Line service types

You can sign up for a variety of line services. They break down into two categories:

- **Analog service**
  Plain old telephone service, Foreign Exchange, Off-premises extension, remote line. Analog services are commonly used for voice services, fax, and low-volume, low-speed data transfer.

- **Digital service**
  Dedicated and Switched 56, Integrated Services Digital Network (ISDN), T-1 Carrier. Digital service is commonly used for high-speed data exchange and for multiplexed voice service (using one pair of wires for multiple voice channels).

Analog services

The majority of phone lines today are analog. Without getting too technical, an analog line is one that transmits the signals of your voice in a way that is similar *(analogous)* to your original voice waveform. If you want to send fax or data over an analog line, you need a modem to change the signal from digital to analog. The receiving end needs a modem, too, to change the signal back to analog.

POTS

This stands for *Plain Old Telephone Service*. POTS lines are analog. With a POTS line you can hold a voice conversation, use a modem, send a fax, hook up an answering machine or add on voice mail. Most business and residential lines today are this type.

Foreign Exchange Line (FX)

This is a POTS telephone line provided from a central office that is outside (foreign to) your local exchange area. This can be useful in a couple of ways: lets you provide local call service to your customers, and gives the appearance of doing business in another area (sometimes useful if you work out of your home and don't want your customers to know it). In addition, a foreign exchange number provides dial tone in the foreign area. This can save you money because calls to and from phones in the foreign area are charged as if you were actually located there.

Off-Premises Extension

You can request that your phone number ring in two locations, such as your home and your business. A separate dedicated POTS phone line is run between the two locations. This technology can also be used to provide answering service to a phone line, though call forwarding is the preferred method these days. Installation costs are usually high. There are also restrictions regarding how far apart the two locations may be. The least expensive option is when both locations are served by the same central office. This service is also called Secretarial Line or Half-Tap.

Remote Line

Lets you have a local telephone number in a distant city. Anytime someone calls the remote number, the call is automatically forwarded to your telephone number. A remote line allows you to maintain a working telephone number when you have no physical location in which to terminate normal telephone service. The remote line exists only in the computer at the telephone company's central office. Even though you don't actually have an office (or a real telephone) at the remote location, you can list the remote number in the White and Yellow Pages.

Using Remote Line (also called Remote Call Forwarding or Market Expansion Line), you can keep a number you have used for business even though you have moved out of the area. Or you can create the appearance of a branch office (or several offices) in other cities. For example, a talent agent located in Fresno uses this service to have a Hollywood telephone number.

You pay for all local and toll charges from the remote number to the terminating number. This means that your caller only pays the local call charge and you pay the direct-dial long-distance charge.

A remote line differs from a foreign exchange line in two ways: 1) a foreign exchange line is more expensive than a remote line, and 2) you can dial out from a foreign exchange line but not from a remote line.

Digital services

Nowadays, more and more organizations are changing to digital lines. Digital is the preferred medium for computerized telephone applications such as data transfer and online communications. Digital lines can transmit at higher speeds than analog, allowing you to send more information in less time, thus saving phone charges. Digital lines are cleaner, therefore providing better sound quality for voice conversations.

Digital will allow multiple telecommunications channels over a single twisted pair copper line. For example, your phone company could expand your one digital circuit into 24 circuits by using multiplexing technology. This would let you have 24 voice phone calls simultaneously over one phone line. Or you could use 18 lines for voice, four for fax and two for data modems—you get the idea. See *T-1 Service* later in this chapter.

Unfortunately, analog telephone devices are not compatible with digital lines. You must use telephone equipment specifically designed for digital lines or get a converter.

Dedicated 56 service
A high-speed digital line that runs only from point A to point B. Used mostly by larger organizations for modemless dedicated digital transmissions like electronic funds transfer, automated teller machines, airline reservations, point-of-sale operations, and data networking among offices or remote sites. This service is also called DDS, ADN, and Leased Line.

You pay a high monthly fee for this service but pay no additional usage charges. You can make unlimited calls at a flat rate. The rate is a combination of service charges and a mileage fee and varies considerably.

Switched 56 service
If you don't need a dedicated line for digital transmission, but still want speed, consider Switched 56 service. This service transmits digital data at 56 Kbps for the price of a voice phone call. Used for high-speed bulk file transfer, distance learning, video surveillance, electronic publishing, video conferencing, shared library transfer, and image transfers (such as advertising layouts, real estate listings, and digitized fingerprints). Allows faster faxing with Group IV fax (seven to nine times faster than current Group III fax), but you can only fax to other Group IV machines. To do this, you replace your modem with a device called a Channel Service Unit/Data Service Unit (CSU/DSU). It connects your computer equipment to the telephone network, makes and answers calls, and manages the flow of data during the call.

Switched 56 rates vary depending on who provides the service, which is available from Regional Bell Operating Companies (RBOCs) as well as long-distance carriers. You'll pay monthly access charges as well as usage charges for this service. Usage charges are usually based on voice rates plus, in many areas, a surcharge ranging from 3¢ to 14¢ a minute.

ISDN service

When people talk about the information superhighway and the amazing capabilities you can get from advanced telecommunications, they're often talking about the Integrated Services Digital Network or ISDN. This technology lets you turn a normal twisted-pair copper phone line into three very fast digital channels (virtual phone lines). Two of the channels are 64 Kbps channels; the third, used for packet-switched transmission (short bursts of encoded information that send information such as Caller ID), ranges from 8 to 16 Kbps. ISDN service lets you combine voice and data traffic on the same phone line. Use one channel as a normal voice line and the other for Group IV fax. Or combine the two channels for high-speed Internet access. All with no new wiring.

Things to do with an ISDN line

• A gas station uses ISDN to run its credit card authorization line over the smaller channel, and still have two voice lines—one for the station's direct line, the other for a pay phone. According to Pacific Bell, tests conducted by major credit card companies show ISDN access to credit card data cuts response time from the 12 to 40 seconds required for an ordinary connection to four to seven seconds via ISDN.

• Realtors are using ISDN-enhanced multiple listing services for fast access to full-screen color images of available properties.

• The local school district in Huntsville, Alabama, employs ISDN to provide a security system that monitors voices and sounds in every school in the city.

• A Hollywood-based music producer sends digitized music across the world at break-neck speeds, using Dolby Fax and an ISDN line.

• Doctors in Palo Alto, California, use ISDN to provide telemedicine links that let them view X-rays remotely.

- Freelancers upload stories, photos and layouts and download edited copy each week for the *San Francisco Examiner Magazine.*

- Telecommuting scientists and engineers at Lawrence Livermore National Laboratories in California use ISDN for remote access to their office LANs (local area networks and file servers.

- A sports photo agency offers an online catalog of over 10,000 images.

- Executive search firms and headhunters conduct virtual job interviews using ISDN videoconferencing.

Speedy downloads

Sharon Crawford installed ISDN service to get faster access to the Internet. Crawford, author of 18 Windows and computer-related books, needed quicker speeds than her 28.8 Kbps modem provided. So she ordered an ISDN line from Pacific Bell and linked it to her computer via an Ascend Pipeline P50 router. She uses a router because she wanted to connect to her home network, where she shares access with her husband and frequent co-author, Charlie Russel.

Their ISDN phone bill averages $80 per month, ISDN has proved invaluable for accessing e-mail, newsgroups, and the Web. While working on a book about Windows 95, she had to download weekly updates of the beta version of the software. "Microsoft set up a private site from which testers could download the new version. We needed to download a 30+ megabyte file about once a week," Crawford says. "This took many hours with a conventional modem and only 40 to 45 minutes with ISDN."

Crawford offers this advice when setting up an ISDN Internet connection: "Get a modem or router that your Internet Service Provider (ISP) is willing to support. If you use a non-recommended brand, the ISP will blame all problems on your hardware (and you'll be unable to prove otherwise.)"

ISDN flexibility

With an ISDN line connected to your PC, fax and phone, you can send and receive just about anything—including documents, full-motion video, voice—and at very high speeds. And speed equates to cost savings.

For example, compare the cost of sending a three-page document from California to Japan using Group III fax with analog lines and Group IV fax over ISDN:

Fax technology	Time	Cost
Group III fax - analog	120 seconds	$8.79
Group IV fax- ISDN	45 seconds	$3.65

Most ISDN service is capable of dynamic bandwidth allocation. This is a fancy way of saying that you can make a data call using the entire bandwidth and, if another call comes in, the system automatically shrinks the first call bandwidth so it can handle the second. For example, you could combine both channels for fast 128 Kbps Web-page turning. Then, when a fax call comes in, the system would cut your Web access speed to 64 Kbps, without breaking your Internet connection, and handle the fax call. Once the fax transmission is over, your 128 Kbps connection is restored automatically.

ISDN service also captures Caller ID information (if it's available). If you have an intelligent ISDN phone, you can program it to switch calls based on called or calling number to certain peripherals (e.g., all faxes to the fax machine and all 800 calls to line one), and avoid the need for a switching device.

Types of ISDN equipment

You'll need some special equipment to take advantage of ISDN's bandwidth. Depending on your application, you may need:

Network Termination Device (NT1)

Sits between your ISDN phone line and the rest of your equipment and serves as your network interface. Some boards and adapters come with NT1.

ISDN router

Allows you to share ISDN service over a network. Try to get a router with an analog port—that way you can use the line more flexibly and plug in a phone or fax machine.

ISDN modem or adapter

Serves as a protocol converter that allows you to attach non-ISDN compatible equipment to ISDN lines. Some come with a telephone jack or two for plugging in an ordinary analog telephone or another device such as a fax, analog modem, standard telephone or answering machine. Most ISDN adapters come with built-in NT1s so you don't have to buy an extra device.

Adapters come in two varieties: internal (installed inside your PC) or external. I like the external kinds better because they have status lights, which come in mighty handy if you need to troubleshoot your equipment.

ISDN telephone

Designed to take full advantage of ISDN's capabilities. Has a built-in terminal adapter and usually an LCD screen for messaging, Caller ID information and other features.

Note: Because ISDN phones require an external power source, you should keep a POTS line just in case of power failure.

Multifunction ISDN device

Jetstream (888-JETSTREAM) makes a nifty ISDN call- management tool called the Front Desk. This amazing device lets you handle up to 16 separate phone numbers over one ISDN line. You can have multiple telephone numbers feed into that one line. For example, you could give one set of clients your "ordinary" fax number and another your "priority" fax number. If two faxes come in at the same time, the device stores the ordinary fax, and puts the priority fax through to your fax machine. And you could still be conducting a data or voice call at the same time as the fax transmission. All for the price of a single ISDN phone line.

Front Desk supports two simultaneous voice, fax, and/or data calls; dynamically routes calls based on time of day, Caller ID, or ring-no-answer; sets up special call screening scenarios that let you get work done and still let important calls go through; handles your voice mail, paging, and faxes; and lets you plug in up to three analog devices. Nice!

Note: Although you need special equipment to take advantage of ISDN's capabilities, you can connect to any number. If you're using ISDN as a voice line, the called party does not need to be ISDN-enabled nor do they have to have digital lines. To achieve ISDN speed, however, you'll want to connect to a fast line on the other end.

You can configure ISDN in several ways. Among the most common for small businesses are:

- **Single line - mixed equipment**
  If you use a "digital modem" such as Motorola's BitSurfr or 3Com's Impact IQ, you can attach your regular telephones and fax machine on to the ISDN channel.

- **Single line - all ISDN**
  All telephones and fax machines must be ISDN-compatible or you can use a special terminal adapter. If you plan to go modemless, your PC must have an ISDN card.

- **ISDN as your second line**
  Leave your voice line as a standard POTS line; put ISDN on the second line.

Ordering ISDN
When ordering an ISDN line, you'll need to know the following:

- What hardware will you be using? Not all hardware will work on the ISDN lines that are available in your area. Much of it is switch-dependent.

- Do you plan to combine the two ISDN channels? If so, you'll want to order a BRI line that is capable of inverse multiplexing, multilinking or bonding.

- How many telephone numbers do you want? If you plan to use your ISDN line with your fax, analog phone, or maybe with a smart device like the Front Desk from Jetstream, you'll definitely want at least two.

So before you go out and buy equipment, especially if you tend to buy from a mail order firm, do some research. And it makes sense to buy the equipment before ordering the line because most vendors provide ordering information that will make your installation much easier.

Preparing for installation

Before you install your equipment, assemble the following information:

- Your Central Office switch type. Your telephone company will tell you what type you have when you sign up for ISDN. Just don't lose this information—it's critical.

- The telephone number or numbers assigned to your ISDN line. You can have up to 16 different telephone numbers, depending on your equipment and your telephone company's capabilities. The most common configuration is two numbers. You'll also need to know how you plan to configure your lines. For example, you might use line 1 for Internet calls and line 2 for fax calls.

- The Service Profile Identification numbers that are assigned to each number. These identifiers tell the central office switch how to talk to your ISDN terminal. Sometimes, you'll hear these numbers referred to as SPIDs.

- It's also a very good idea to have the equipment vendor's customer service phone number handy.

Note: ISDN is still not plug and play. If you're not technical and don't have time to mess with all this, get help from your vendor or a consultant to set it up. See Resources at the end of this chapter for some useful contacts.

Costs

ISDN prices vary significantly depending on where you live. Installation costs range from $35 up to $250. Monthly rates vary also—from $33 to $105 depending on where you're located. Most companies also charge a low (1 - 3¢) per-minute usage charge. For more information, see *Resources* at the end of the chapter.

ISDN has been around for several years and is quite common in Europe and Japan. It is only now reaching prices that make it attractive to American home-based businesses and small offices. And it's still not available everywhere. But it will be.

T-1 service

A T-1 line provides point-to-point digital transport capacity of 1.544 Mbps. You can multiplex the circuit (add channels to a single pair of wires) in a variety of ways—one of the most common turns your two-pair copper phone line into 24 voice channels. T-1 doesn't come cheap. Each month you pay for a local channel, an access connection fee, access coordination, multiplexing charges, as well as mileage charges. You also need special equipment at each end of the circuit. According to industry sources, a standard 24-channel connection will cost about $2,800 in fixed monthly charges plus $3.95 per mile. The break-even point, when comparing it to simply adding separate lines, is at about 12 lines.

T-1 is not cost-effective unless you use it three to six hours a day. If you have heavy digital transport needs and need more capacity than ISDN can offer, but don't need a full T-1 bandwidth, you can arrange for Fractional T-1 service in multiples of 64 Kbps.

Speed comparison

It's helpful to think of transmission speed in terms of lanes on a highway.

1 lane (or less)

A POTS (Plain Old Telephone Service) analog line, the kind most of us use, can't even get on the highway without a modem. The fastest modems you can get for an analog line top out at 56 Kbps (kilobits per second). If the 56 Kbps modem represents one lane, a 28.8 modem is about one-half of a lane. A 14.4 modem would give you one-quarter of a lane. A 9.6 modem would give you one-sixth of a lane.

1 lane for sure

Dedicated or Switched 56 service transmits at 56 Kbps—the equivalent of a one-lane road.

2 lanes

An ISDN BRI line more than doubles that capacity and can give you up to 128 Kbps. ISDN gives you two lanes.

24 lanes

T-1 runs at 1.544 Mbps and is the equivalent of 12 ISDN lines, or 24 lanes. If you think that's fast, fiber-optics can provide 32,000 lanes.

On the horizon - ADSL

Here's another acronym you'll be hearing lots about in the near future: ADSL. It stands for asymmetric digital subscriber line, and is a new service just rolling out in select areas. ADSL has downstream speeds of 1.5 to 9 Mbps and upstream speeds ranging from 16 to 640 Kbps. In plain English, you'll be able to receive data at up to 1.5 million bits per second (about 50 times faster than a 28.8 modem) and send data at 64,000 bits per second (twice the speed of a 28.8 modem).

What's neat about ADSL is that it also runs over copper phone lines—no need to rewire. What's not so neat is that ADSL will not be cheap. Line costs range from $50 to $125 a month. Add to that an additional Internet monthly charge of about $65 a month and then you'll need to buy an ADSL modem for around $500. And, unlike ISDN, you can't use an ADSL line to access several different destination lines. It has to be set up to call the same phone number every time (for example, your Internet Service Provider). ADSL also has certain distance limitations—check with your phone company to see whether it will work for you.

Central office secrets

Ever hear of the term "G.U."? It's short for Geographically Undesirable and refers to a guy or girl who lived too far away to be considered appropriate date material. Well, your local phone service may make a particular business location G.U. Here's how:

Local phone service comes into your home or office from the telephone company's central office. You are assigned a telephone number with an exchange prefix (the first three digits of your seven digit telephone number). There is a physical limitation on the quantity of phone numbers available from an individual exchange. This limitation is set by the switch type (ranging from 30,000 to 130,000 phone numbers). Smaller central offices may have just one exchange prefix per office, larger ones may have several. So what, you say? Well, getting the wrong exchange may do serious harm to your business communications plans. Why? Older telephone exchanges cannot provide the calling features you may need.

- A photographer wanted to add Call Waiting to her phone service. When she called her phone company, she found that Call Waiting was not available on her exchange because the switch did not contain the proper electronics to service it. The only solution was to change exchanges, which involved changing her telephone number, reprinting business stationery and notifying all her clients—a very costly proposition. She was lucky, in a way, because her central office had more than one exchange with Call Waiting available. If her central office had only one exchange, the only way she could get Call Waiting would be by paying for an expensive foreign exchange line. Or she could move across town.

- An architect, specializing in CAD/CAM renderings, wanted to relocate to a country setting. Since the majority of his work was done via telecommuting, he felt he could basically settle anywhere in the country. All he needed was a telephone line. He found an idyllic spot and set up shop. Things were fine until he wanted to expand. He needed to add a third line to his setup and wanted to add Call Hunting capability (this groups telephone numbers so that if the first line is busy, the next line in the hunt group rings, and so on). Unfortunately, his central office had only one exchange and that exchange could not support Hunting unless the numbers were physically located next to each other in the switch. And the numbers needed were assigned years ago. His tough luck.

- A small manufacturer lost a major contract because the company wanted all its vendors to support EDI (electronic data interchange) and that required digital service. When the manufacturer checked with his local telco, he learned that his local central office had an earlier version ESS (Electronic Switch System) switch, which was not digital. Sorry.

And woe betide those of you that live in lovely remote areas far from the Central Office. Inexpensive services such as Voice Mail can become very costly, when you have to pay mileage rates.

When you are considering a business location, find out what type of phone switch your telephone company provides for the exchange. Then call the local phone company, give them the address of the property you are considering, and ask for the type of switch. If the switch is a stepper or crossbar exchange, you can't do much more than add touch-tone dialing. You want an ESS, or better yet, a DMS (Digital Multiplex System) switch.

If the service agent doesn't know the switch type, ask what kinds of services are available on the exchange. If the list of services is limited (e.g., no Call Waiting or Call Forwarding features), you're out of luck. Additionally, if you plan to use advanced communications services like ISDN, be sure to find out if they're available in your area, or how soon they will be. You'll be glad you did your homework.

Ordering service
When you're ready to order a new line, prepare yourself before you call your business office. The types of services available and the rates you pay for local service are determined by your class of service (CS). Often the class will have different features and services associated with it.

- **Residential** - If your office is in your home, you can often qualify for a residential CS. These rates are often considerably lower than business rates.

- **Business** - A business CS has a higher rate than a residential rate but provides a free listing in the Yellow Pages.

Payment plans
Most telephone companies offer a variety of payment plans. These are the most common:

- **Measured service** - You pay a lower monthly access charge, but each local call is billed. The cost of the call is often variable by time of day. Some measured rate service is untimed, but the majority is timed—the shorter and fewer the calls, the less you pay.

- **Flat-rate service** - Some areas of the country allow you to pay a fixed monthly sum for an unlimited number of local calls. This is usually available for residential services only. If this is available in your area and you can qualify for residential service, get it—it's a bargain.

- **Party line service** - Believe it or not, there are still some areas of the country that offer party-line service. This type of service groups two or more separate subscribers on one line. It's usually available in remote areas of the country where the cost of stringing copper wire makes dedicated service prohibitively expensive. If you find your business located in such an area and you don't yet have service, consider moving or look into wireless services (see Chapter 7).

Other factors
In addition to the services and features you want on your line and your class of service, you'll also need to consider other factors:

- **Touch-Tone** - Yes, you've got to order Touch-Tone service. In many areas, you also have to pay a small monthly fee for the service. You should order Touch-Tone—without it, you can't take advantage of powerful features like on-demand Call Forwarding and voice mail.

- **Your new number** - You have some control over the number you get. If you want a particular number like 355-SAVE or 4APIZZA, ask your phone company if that number is taken. The phone company may charge you a research fee and a monthly charge for the vanity phone number. Still, it might be worth the extra cost.

    Hint: Take a look at the PhoNETic Web site *(www.soc.qc. edu/phonetic)*. This site provides an interactive tool that lets you test out different combinations by converting letters to numbers.

- **Inside wire maintenance** - For a low monthly fee (often around 50¢ a line), you can contract with the phone company to provide you with a monthly repair service plan. Under this plan, phone company repair persons will visit your site when you report a phone problem and, if the problem turns out to be faulty wiring inside your facility, will repair the line at no extra charge. If the lines are fine and the problem is caused by your equipment, they will not charge you for a "needless visit."

    You might not need this service. According to the authors of *The Complete Guide to Local & Long Distance Telephone Company Billing*, inside wire breaks on average once every 14 years. On the other hand, just one mistake with inside wire will cost you a minimum of $35 for an onsite visit—and that's a lot of wire maintenance payments.

- **Your listing** - If you have business service, you're entitled to a free listing in the business White Pages and, in many parts of the country, also to a free listing in the local Yellow Pages. You might want to investigate the cost of a display ad or at least boldface type. Check out your competitor's ads. Consider listing in more than one phone book. If the bulk of your business is by phone, it makes sense to increase your exposure.

    For advice on writing an ad, take a look at *Getting The Most From Your Yellow Pages Advertising*, 2nd edition, by Barry Maher (Aegis, 1997).

Resources

Getting The Most From Your Yellow Pages Advertising
2nd edition
by Barry Maher
Aegis, 1997

ISDN - books, guides, newsletters
National ISDN Users Forum
Maintains a catalog of ISDN applications that you can download via the Internet at info.belcore.com or call Bellcore (800-521-2673) and ask for Document # GP-1

The ISDN User Newsletter
Information Gatekeepers Inc.
Boston, MA, 1995
800-323-1088

ISDN: A User's Guide to Services, Applications & Resources in California
Pacific Bell, 1994
2600 Camino Ramon
San Ramon, CA 94583
800-472-4736

1995 International ISDN Yellow Pages, Third edition
Information Gatekeepers Inc.
Boston, MA, 1995
800-323-1088

ISDN: How to Get a High-Speed Connection to the Internet
by Charles Summers and Bryant Dunetz
John Wiley & Sons, 1996

ISDN information by phone
Bellcore ISDN HotLine
800-992-4736

ISDN information on the Internet
Dan Kegel's ISDN page: http://www.alumni.caltech.edu/~dank/isdn/

ISDN InfoCentre: http://www.isdn.ocn.com

Usegroup News: comp.dcom.isdn

5...Stand-Alone Phones

Not too long ago, if you needed a phone, the most important decision you'd need to make was whether the color should be black, white or ivory. Later, feature phones added Touch-Tone capability, memory dialing, redial and the like. Now phones come in a range of styles with hundreds of features—speaker-phones, cordless phones, conference phones, fashion phones, clock-radio phones, PC-phones, videophones—even a wrist-phone (shades of Dick Tracy!).

A phone may either be stand-alone, which means that the phone needs only a phone line to operate; or the phone can be part of a phone system such as a key system or PBX. Phones also differ in how they communicate. Landline phones use telephone wire to send their signals; cellular phones use airwaves. Cordless phones use a combination of landlines and airwaves.

This chapter is all about stand-alone landline phones. I'll discuss how they work, what all those different features do, and what to look for in a single, multi-line telephones or cordless model. There's advice about troubleshooting and information about selecting and using a headset. Finally, there's a section on new telephone technology, including phones that operate through your personal computer. Chapter 6 will cover phone systems and Chapter 7 will discuss wireless phones.

How phones work

When you were a kid, did you ever build a play telephone using two tin cans connected by a long string? Your friend spoke in one can and you could hear from the other? Guess what? Phones work like that, sorta.

The handset of a modern telephone contains a microphone in the mouthpiece to amplify the sound of your voice. As you speak into the mouthpiece, your voice causes air vibrations. These vibrations cause a membrane, stretched across carbon granules in the mouthpiece, to move. This creates resistance that causes the current that contains your voice waves to flow down the telephone wire. The earpiece, which may or may not be connected to the handset, contains another vibrating diaphragm and a tiny loudspeaker.

When you dial a telephone number, the instructions are sent through a network of exchanges, finally resulting in ringing the phone on the other end. When that phone is picked up, the two handsets are electrically connected. The sound your caller hears is your voice converted into a series of electrical signals. The cracks, pops and other weird sounds you sometimes hear are electrical interference on the line. With digital lines, you don't experience all that interference.

Interestingly, the ring sequences you hear while waiting for the call to be picked up do not match the ring sequences heard by the dialed party. Early phones just had dead silence until the caller picked up. This drove people nuts so phone companies now add ringing tone on the calling party's line. The rings you hear are put there by your phone company so you'll feel satisfied that the call is being processed. The rings at the other end are often one or two rings fewer than the number of rings you heard. So, be patient—six rings to you may only be four rings on the other end.

Single-line corded phones

You can get a simple single-line corded phone with nary an extra button for less than $20 or spend over a hundred for a phone crammed with features. Phone manufacturers are adding an incredible array of functions these days. Is it progress or creeping featuritis? You decide.

Phone features - dialing

Alphabetic keypad

In addition to the standard numeric keypad, some phones provide a separate 26-key alphabetic keyboard. Handy if you do a lot of phone programming.

Automatic redial

Automatically dials the last number you dialed. Some phone models will keep trying if the line was busy. *Note:* this is also called Last Number Redial.

Chain dialing

This automatically dials a sequence of numbers with appropriate pauses. Useful for bank-by-phone, voice mail and other applications that require a sequence of numbers. For example, you dial your bank's phone number and key in your account number after they answer.

Dialing mode selector

Lets you change from Touch-Tone to pulse dialing. Useful if your local service doesn't support Touch-Tone and you need tones to access a voice mail system or other computerized services. To do this, you dial the service with pulse dialing, then you switch over to tone to communicate properly with the call processing computer.

Emergency dialing

Some phones come with pre-labeled buttons for fire, police and ambulance services.

On-hook dialing

Lets you dial without having to pick up the handset. If you use a headset, you'll definitely want this feature because it saves you the trouble of picking up the handset every time you make a call.

One-touch save

Saves the last number you just dialed in a scratch pad. Number will be erased the next time you save a number.

Saves time if you have to dial the same number several times during a particular day.

Programmable soft keys

Lets you control which features to assign to certain keys. For example, you might program one key for last-number redial and another for three-way conferencing.

Speed dialing

This is one of the more popular features. You program frequently-called numbers and can dial them with a one- two- or three-button code. Make sure that the phone's memory can handle the longest number you want to store, including country code, area code and long-distance access number. Eleven digits is a bare minimum these days. If you dial internationally, you may need 16-digit capacity. *Note*: this is sometimes called One-Touch Speed Dialing.

Voice activated dialing

Some advanced phones let you program a list of numbers with a voice tag. Then, when you want to place an outgoing call, you just speak the name: "delivery service" or "taxi."

Voice response dialing

Your phone repeats the number you dialed (in English or Spanish) as you dial. Could be useful for the dialing-challenged.

Phone features - display

Call forwarding light

Reminds you that you forwarded your phone calls. You'll need a corresponding service from your phone company for this one to work.

Caller ID compatible

Displays the local caller's telephone number (and sometimes, the caller's name) if the information is available. You must sign up with your telephone company to get

Caller ID service. Some phones also have memory capability for storing a running list of incoming caller's numbers. This memory feature even records the phone numbers of callers who hung up before you answered.

LCD display

Shows the number being dialed, the date and time. Some phones also display the elapsed time of the call, useful if you need to bill for time on the phone or watch those long-distance minutes.

Visual ring indicator

Lights up when the phone rings and allows you to know that a call is ringing in. Especially useful if you are hearing-disabled or use the phone in very noisy conditions.

Phone features - control

Busy supervision

If the number you dialed is busy, some phones will automatically hang up for you.

Feature keys

Gives you one-button access to phone company features such as Call Return, Repeat Dial, Three-Way Calling, Call Forwarding and the like.

Flash

This is a useful button for those of you with phone company services like Call Waiting or Three-Way Calling. Pressing FLASH is just like pressing hookswitch except that you can't accidentally hang up on your caller using FLASH. If you have a tendency to be heavy fingered, this is definitely a useful feature.

Hold

Lets you keep the caller on the line, but you can't hear him and he can't hear you. Especially thoughtful to the caller if your workspace is noisy.

Intercom

Lets you talk with other extensions in your facility. If coworkers or family members occasionally answer your line, you'll want intercom capability. That way they can call you to the phone by ringing your extension.

Message waiting light

Lights up if a call came in and went unanswered. Especially useful if you have phone company voice mail.

Mute

Press the MUTE button and you can still hear your caller but your caller can't hear you. This can be handy if you are working at home or in a very noisy setting and you want to ask your coworkers (or kids) to quiet down. Also handy if you can't stop a sneeze.

Security

You can set up a security code that locks your speed dial directory and prevents outgoing calls (other than to emergency numbers).

Phone features - comfort

Handset volume control

Lets you adjust the volume of each caller's voice. Nice if you're stuck with a shouter.

Hearing-aid compatible

Because a hearing aid is a microphone and your phone earpiece is a speaker, putting the two near each other can cause painful electronic feedback. A hearing-aid compatible handset has circuitry that cancels the feedback.

Ringer

You can get all kinds of ringers. Some phones quack, others chirp, some play music instead of the standard ring. There's even one that tells you, in English, that you have a phone call. Whatever ringer you get, you'll want to control the volume and maybe even turn it off.

Speakerphone

Phone has a built-in speaker that lets you talk without having to hold the handset next to your ear. The built-in microphone is omnidirectional so that it can pick up your voice wherever you may be in relationship to the phone. This type of mike also picks up every other sound in the room, including the echoes your voice generates as it bounces around the walls. It's that echo effect that is so annoying to the person at the other end of the connection. You sound as though you are speaking from the bottom of a well. To improve your speakerphone's sound quality, use it in a room with lots of cushioning such as carpets, drapes, and books.

Speakerphones aren't private. The called party has no idea, unless you clue her in, of who else is within hearing distance of your conversation. This, combined with the echoing sound quality, can definitely put people off.

I use my speakerphone for dialing or when I'm placed on hold. That way, I can keep on working until someone comes on the line. When that happens, I turn off the speakerphone and use the handset. On rare occasions, I continue using the speakerphone during the conversation if I need to take copious notes, and then only with permission. I also use it with automated voice response systems while listening to a menu of voice choices.

Almost all speakerphones are half-duplex. This means that only one party can talk at a time without clipping off the other person's voice. If you need speakerphone capability for conference calling, get a full-duplex phone, or consider getting a conferencing system, discussed later in this chapter.

Multi-line phones

Multi-line phones allow you to juggle multiple phone calls, place a caller on hold and dial out on the other line to get needed information, or even conference callers together.

If you have more than one voice line coming in to your workplace, you should invest in a multi-line phone. The alternative—a desktop arrayed with single-line phone sets and the guessing game to determine which line is ringing—is a real time-waster. Multi-line phones come in two-, three-, four-, and even five-line models. Depending on your wiring config-uration, you can have just one line cord between your phone and the wall jack for each two lines. (*Note*: most homes and offices are already wired for two lines.) Once you've reached five lines (or even before), you're ready to graduate to a phone system—a key set, hybrid, PBX or Centrex. (See Chapter 6 for details.)

I have a two-line phone in my home office. I use line one for incoming voice calls and line two for outgoing calls as well as fax and modem calls. A single line cord (called a 2-line or RJ-14 cord) runs from the phone to the wall jack.

Multi-line features

Multi-line phones come with many of the features of a single line phone. In addition, they have special features designed for handling more than one phone line. Some of those fea-tures are:

Conference button

You can connect the caller on the first line with the caller on the second line by pressing the CONFERENCE button. This is a very useful feature—lets you have a three-way conference call without paying extra for the phone service kind.

Do not disturb mode

Makes your phone busy without being off-hook. This is useful when you need an uninterrupted chunk of time to complete a project or are having an important conference that can't be disturbed.

Intercom capability

Some models allow you to page more than one phone set at the same time.

Line buttons and status lights

Of course, you'll need to see which phone lines are in use. All multi-line phones have similar methods for selecting a line and have lights or a display panel that indicate active lines, lines on hold, and on which line a call is ringing.

Number of lines supported

The most common multi-line phone supports two lines. These phones usually have two incoming jacks: one for line one and one for lines one and two. This allows you to use the two-line phone as a single line instrument. Three-line phones may have two jacks—the first jack gets a single-line connection; the other jack gets lines two and three. Alternatively, if your wiring supports a three-line connection, you can attach just one three-line cord. The two jacks that support four-line phones each have a dual-line connection. Five-line phones use a two-line connection for one jack and a three-line connection for the other.

Ringer control

You'll want to be able to selectively turn off the ringer on one or more lines, especially if you have that line backed up with voice mail or an answering device.

Speed dial portability

Lets you transfer the contents of your speed dial directory to another phone set. Convenient if you need to share a common set of speed dial numbers with others in your organization.

Cordless phones

If you need to walk while you talk, consider getting a cordless phone. With a cordless phone, you can get up from your work area, pull files, answer the door, walk outside and still be connected. Though earlier cordless models were noted for their poor audio quality, most models today operate at or near corded phone quality.

Cordless phones are a combination of wired and wireless technology and come in two parts:

•   **Base station** - this part plugs into a telephone jack and an electrical power outlet and stays put.

- **Portable handset** - contains the mouthpiece, ear piece and battery. A few cordless models come with an optional headset.

The handset communicates with the base station over radio waves. Both sides of the conversation are beamed between the handset and the base station. Until recently, cordless phones were limited to only 10 frequency channels. With so few channels, the chances that a neighboring phone was using the same frequency channel was high. When that happened, you could overhear their phone conversation, and they were able to hear *yours*. Your only recourse was to switch channels. If all the channels were in use, you just had to wait. Now, higher-end cordless phones support over 100 channels, reducing the possibility of channel contention.

Ever hear a crying child on your cordless? If you have a cordless phone in your home office, and you or a neighbor has a baby monitoring system, don't be surprised if you hear a baby crying over the phone some day. Room monitors and cordless phones operate in the same frequencies.

Earlier cordless phones had a major problem: Anyone with any handset could dial out over your base station and place expensive long-distance calls that would be billed to you. Happily, phone manufacturers have curtailed this fraud by adding encoded digital security systems.

Cordless phones have a restricted operating range. Those with 46/49 MHz handle ranges from 300 to 1,000 feet. The more expensive 900 MHz phones claim a range of up to a half a mile. The ranges quoted by manufacturers are optimum figures. Your actual range will be considerably shorter because it is affected by walls, metal shelving and interference from other electrical appliances such as PCs and fax machines.

The communication between the handset and the base depends on battery power. The batteries recharge when you replace the handset in the base station. If you forget to replace the handset, your batteries will begin to lose their power and, as they fade, the effective range of the phone will be affected.

Unlike a corded phone, you have to turn a cordless phone on to place or receive calls. Cordless phones will not work if there is a power outage, so don't rely on a cordless phone as your only phone.

Because a cordless phone is so portable, it's easy to leave it lying around wherever you finished your last phone call. Try to avoid this. It's not much fun hunting madly for the handset when the phone rings. Invest in a model that has two-way paging. That way, you can push the PAGING button at the base station, causing the handset to ring and you'll be able to track it down.

Note: if you plan to use more than one cordless phone in your workplace, be sure to check with your vendor to determine how many cordless phones can be used in the same environment and how far apart the base stations must be situated. Also be sure to set each phone to a separate channel.

Cordless features
Because cordless phones share many features with the corded variety, I'll mention only those features that are unique.

Antenna
Flexible antennas are less likely to snap than the telescoping metal kind. A few cordless models use your hand and arm as the antenna. Incidentally, if your antenna is broken, you can buy an inexpensive clip-on replacement at most phone stores.

Auto answer
Lets you receive a call without having to press the TALK button. You just pick up the handset at the base unit.

Audio encryption
Various systems are used to scramble the voice signal, preventing electronic eavesdropping and ensuring privacy. If someone is listening in, all they hear is garbled speech.

Backup power
A few sets come with built-in backup power (an extra charged battery stored in the base unit) which allows phone calls even during a power failure. Crucial, if you have no corded phones.

Battery life

Battery life is usually stated in two modes: talk time and standby time. Talk time between charges can vary considerably, depending on the battery type. Standby times of 14 days are not unusual. Some models come with an extra battery stored in the base that is always charged up and ready to go.

LCD display

Shows number dialed, channel, range and battery status. Some also clock elapsed time of the phone call.

Lockout feature

Prevents phone fraud. Another cordless phone user in the neighborhood cannot dial out using your phone line. See *Security Codes*, below.

Low voltage meter

A visual status indicator of battery life.

Number of channels

Because the channels over which cordless phones may communicate are limited, your phone conversations could be overheard on a neighboring cordless phone. Most cordless phones today offer 10-channel capability. Higher-end 900 MHz phones offer up to 100 channels. You can change channels manually at the handset or let the phone automatically scan to find the clearest channel.

Out-of-range indicator

A beeping tone or alarm lets you know if you are roaming close to the limit of the cordless phone's range.

Paging capability

Lets you ring the other half of the phone. Useful for intercom purposes and also for locating a missing handset. Two-way paging allows you to fully use the intercom capability. Some sets only offer one-way (base to handset) paging.

Radio frequency

Cordless phones come in two choices of frequency: 46/49 MHz and 900 MHz. The 46/49 MHz frequency allows a range of up to 1,000 feet. Because this frequency has a wave length of 18 feet, the effective indoor range is often less than one-fifth the maximum range. This frequency works most effectively in uncluttered open spaces or single rooms. The 900 MHz frequency allows a range of up to a half a mile. It can deal more effectively with interference because its wave length is just one foot. 900 MHz phones are much more expensive than 46/49 MHz phones.

Ringer

A ringer in the base unit is handy, especially if you have a speakerphone. Often you can change the ringer's volume or turn it off.

Security codes

Each phone comes with a number of security codes that help prevent phone fraud. Early models only had a few; today you can get anywhere from 1,000 to 100,000 and up. Most models will randomly change the security code for you.

Speakerphone

Some sets come with a speaker built-in to the base unit. Some speakerphones also have a keypad in the base as well as the handset. This gives you the capabilities of two extensions. You can conduct a three-way conversation between a caller, a coworker and yourself by using the speakerphone and the handset. Some cordless phones don't have fully functional speakerphones and are used only for intercom functions. Be sure to ask.

Troubleshooting tips - cordless

✔ Make sure that your base station is located in a central location and preferably on a high shelf, far away from interference from other electrical appliances.

✔ Raise the antenna on the base unit and the handset for best reception.

✔ Don't plug the base unit into a circuit that also powers a major appliance. To do so will probably cause interference and could greatly limit your range.

✔ Radio interference can be caused by such things as a TV, refrigerator, vacuum cleaner, computer, fax machine, fluorescent light or electrical storm. You can do something about the electrical stuff in your office, but if Mother Nature is active, your best bet is to use a corded phone.

✔ If you experience noise on the line, select a different channel (or let your phone automatically search for a clear channel). If you still experience static, move closer to the base unit.

✔ No dial tone? Could be many things. Is the phone in STANDBY mode? If so, select TALK. Is the phone switched off? Are the batteries low? Is the telephone cord detached from the base unit? Is the base set plugged in to electrical power? Maybe the security codes in the handset and base unit got out of sync. Try placing the handset back on base and see if that corrects the problem. After you've checked all this out, and the phone still doesn't work, it's possible that your phone line is bad (See Chapter 2 for suggestions).

Troubleshooting tips - other phones

Static on the line
Static on corded phones can be caused by many things. Electrical storms, crummy wiring, and damp connectors are among the culprits. If none of these things seem to be at fault, try this old trick: Tap the mouthpiece with the palm of your hand. This shakes up the carbon granules inside the mouthpiece and realigns them. High humidity can make the granules stick together, creating interference.

Read your manual
This really happened to me. One day my two-line phone started going wacky. I first noticed the problem when I called 411 for information and I got *two* information operators. Then, when I called my voice mail provider to get a telephone

number, I was connected to two service agents. But the phone worked fine when calls came in. Strange.

The Panasonic Easa-Phone had been in the family for about three years so, I thought, maybe it was just getting old. Maybe the lines got crossed up somehow and I was dialing out on two lines. I was correct about this, but not so correct about my next step. I thought it must be a low-battery thing. So I dug around, found three okay-looking batteries and swapped them out. No luck. Tried changing line cords. Nope. Replaced the phone with a working single line phone. This worked, but I was now down to only a single line.

Later that day, I called my friend the phone guru and asked for his advice. "You've probably pushed in the conference button," he suggested. "What conference button?" I asked myself.

So finally I turned to the operations manual. And there, in black and white, was a simple drawing explaining the workings of my set. The phone did indeed have a conference button (cryptically labeled CONF) that I just never noticed. In retracing the events leading up to the phone's Jeckyl & Hyde behavior, I recalled that my cat Winston, who was insanely jealous of the time I spent on the phone, jumped on it the previous day and, I guess, landed on the CONF button. This linked the two lines together. All I had to do was push the CONF button again to break the connection.

The moral of this story: Read your phone manual. It contains information that can definitely save troubleshooting time. Or, I suppose I could kick the cats out of the office. Nah!

Headsets

If you spend several hours a day on the phone, or are suffering from headaches and neck cramps caused by scrunching the phone between ear and shoulder, it's time to get a headset. Headsets come in corded and cordless models and let you take calls and work comfortably hands-free. Headsets are being used by more and more professionals—stockbrokers, technical support engineers, travel agents, journalists, pharmacists, marketing and sales personnel, as well as by dispatchers, telemarketers and customer service agents. Prices for headsets range from a low of $50 to around $200 depending on the feature set.

Headset use reduces neck and back strain, adds ergonomic comfort to your workspace and decreases stress and job turnover. Industry studies show that using a headset will increase productivity by at least 11%. If you're on the phone two hours a day, headset use will save you approximately an hour a week.

Productivity increase based on headset use

Time spent on phone calls per day	Hours saved per day	Additional calls possible per day
25%	.22	4
50%	.44	8
75%	.66	12

(Source: "Do Headsets Save Money? You Bet!," *TeleProfessional*, October 1993).

I love my headset phone, a hands-free cordless from Plantronics (800-544-4660). It comes with a light-weight remote pack that I carry in my pocket or suspended around my neck. This lets me roam around the house or yard and answer calls with a single touch. And I really like the ability to type comfortably while on the phone with no neck cramps.

Headsets replace your handset and use a modular plug to connect to an amplifier that plugs into your phone base. Your handset is connected to the amplifier as well. Headsets usually have a flexible boom with a tiny microphone at the end to pick up your voice.

Connecting a Head Set

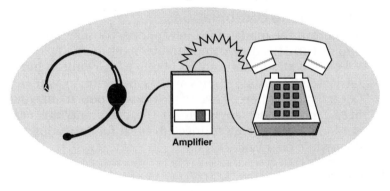

Amplifier

Headset features

Amplifier

The headset connects to the amplifier, and the amplifier connects to the phone. Good amps can increase the volume of the call up to 30%.

Compatibility with your telephone

Headsets are compatible with most business telephones. Check with the headset vendor to make sure.

Cordless capability

Some headsets are cordfree and let you roam as far as 50 or 100 feet from the phone base.

Earpiece(s)

You can get headsets with one earpiece or two. The single earpiece lets you hear what's going on around you. If you work in a noisy or distracting environment, or suffer from hearing problems, get a headset with two earpieces.

Headband or over the ear?

Some models hang over your ear; others fit on a headband. Some headbands are more adjustable than others.

Headset/phone combinations

If you're in the market for a new phone, consider buying a phone with headset compatibility built-in, such as the VTech 900 MZ digital cordless phone with a built-in headset jack. The phone comes with a belt clip so you can "wear" it anywhere. Plug in a headset, pop it on your head, and wander anywhere in the office and the near outdoors and still stay connected. Great for answering the door while on a call!

Microphone

Make sure the microphone has noise-canceling capability. Otherwise, your callers may hear extraneous background noises.

Microphone boom

You'll want a telescoping boom so you can position the mike as close to your mouth as possible. Some headsets have a boom that swivels so you can wear the mike either on your right or left side.

Mute

You'll want mute control. This lets you continue to hear the caller but cancels out your mike. Useful if you can't suppress a sneeze (or a side comment about your caller). Some headsets put mute control in the microphone boom. Up for MUTE; down for TALK. Others place the mute control in the amplifier.

Operating range

If you choose a corded model, make sure the headset cord is long enough to let you reach for a file or turn around without pulling up short. You don't want the feeling of being tethered to your phone. If you go for a cordless model, find out how far you can roam before static interferes (some travel up to 1000 feet).

Quick disconnect

If you opt for a corded headset, you'll want an easy method for disconnecting the headset from the amp or phone so that you can get up and walk around without having to take the headset off.

Y-adapter

Lets another person plug in to your phone line. Useful for monitoring and training purposes.

Weight

The lighter the better. However, lightweight sets may have fewer built-in features.

Earphone

If you want cordless convenience without the hassle of a handset and don't like wearing a headset, look into the Jabra EarPhone (800-EAR-2230). This is a tiny, thimble-sized device that fits comfortably in your ear and replaces your telephone handset. It contains a miniature microphone that picks up only the sound of your voice. The earphone is connected by a cord to a clip-on control unit that contains mute and volume controls. The control unit plugs into the handset jack on your phone. Jabra also makes a model that works with several different cellular phones, and another that turns your Power Macintosh into a smart phone.

Using a headset

Headsets are a bit like glasses and take some getting used to. It's best to wear the headset for short intervals and increase the wear time each day until you are handset-free (should take two to three weeks).

Headsets take the place of handsets in most situations but, unless you buy a special phone designed to work only with a headset or have a LINE or SPEAKERPHONE button, you'll need to lift the receiver in order to get dial tone. The receiver must remain off-hook during the call and be placed back on-hook before you can receive the next call. To simplify things, get an off-hook device such as the Touch-N-Talk from Hello Direct (800-HI-HELLO). This device attaches to your phone base and lets you make a call by pressing a lever which lifts the headset. At the end of the call, just lower the lever.

Another problem with headset use is that it's not always easy for your coworkers to tell when you're on the phone. If you want to reduce interruptions, you might install a Busy light, an add-on device that lights up when you're on the line.

Phone accessories

Amplified phones and handsets

If you suffer from a hearing loss or the inability to hear high-frequency sounds, you might consider investing in an amplified phone. You can get one from Hello Direct or Lucent Technologies. These phones do more than let you adjust the volume on the earpiece. They contain circuitry that improves the clarity of the sound by selectively increasing the volume of only high-frequency sounds. Controls for tone and volume settings are usually found in the handset.

Another alternative is to get a handset amplifier. These come in several variations. One type is plugged in between your handset cord and your handset and has a volume control dial. Another type replaces the regular handset on your telephone with a special handset with a volume control built in.

Recording devices

Writers, attorneys, real estate agents and many others who need to keep an accurate record of their calls find phone recording to be a useful tool. Recording lets you avoid the difficulty of trying to take copious notes while on a call.

Some people think that you must have a device that generates an annoying beep-tone every 15 seconds or so while recording. This is not required. According to the FCC (Rule 6, subpart E), you may legally record a phone conversation as long as you inform the other party that you are recording. As an extra safeguard, record the notification at the beginning of the recorded conversation.

If you need to record a telephone conversation, you can get a voice-activated recording adapter. This has a modular jack on one end for plugging into your phone line (some plug in between the base and the handset) and a plug on the other to connect to your tape recorder. Then, just turn on the device to tape a call. The recorder will pause when there's no voice on the line.

Call signalers

Can't hear the phone ring over the sounds in your workspace? Need to hear the phone ring from great distances? Want to know if the phone is ringing but not have to hear it? There are products to solve every phone ringer issue you can think up. Shop around for a call signaling device. You can get an adapter that turns off your ringer and activates a strobe light; one that augments your ringer with a loud horn, warble or tweet; and even one that makes your desk lamp flash when a call comes in.

Conferencers

This is for those of you that need to conference several people on a call, and need to do it frequently. A conferencing system is a fancy speakerphone that has several built in microphones for 360° sound pickup and digital signal processors to cancel out voice echoes. The speakerphone is full-duplex so that you and your called party can talk at the same time without experiencing the clipping and one-way conversation quality of standard speakerphones.

What's new?

Wristphone

Ever wanted to emulate the comic book heroes and make a call right from your wrist? Now you can. MicroTalk Technologies (612-545-2627) offers the TeleWatch, a combination watch, cordless phone and office intercom unit. The watch can make and receive calls within 300 feet of its base station.

The everything phone

Navitel makes the TouchPhone, a multifunction device that allows you to browse the Internet via a 6-inch screen, access Caller ID information, set up automated address books, see who's calling even if you're on the phone (with Call Waiting Caller ID), access your calendar and scheduler, see an in-box with all your voice mail and e-mail messages, and forward select calls. This phone is also a digital answering machine with over 20 minutes of recording time. You also get a calculator, alarm clock and two-line telephone capability. You can control the device using a touch screen or via a built-in typewriter keyboard that hides away in a drawer when not needed. For information, contact Navitel at 415-462-9171 or *www.navitel.com.*

Portable videophone

Casio (800-HELP-134) makes a portable videophone small enough to fit in your briefcase. The LT-70P Videophone works with any VCR-compatible TV and uses a single normal copper phone line. To set it up, you connect the Videophone to a telephone through the modular jack on the back of the phone, connect the phone to a TV using the supplied AV cables and plug the whole unit into power and phone lines. Aim the built-in camera at the image you want to transmit, place a call to someone who has a similar unit, and talk and watch to your heart's content.

PC phone

Take a computer, plug in a telephone line, add some software, a microphone, and maybe an internal card, and you get a very smart telephone. Computer telephones save valuable desktop real estate and provide greater functionality. Developers are creating computer software that runs on Macs or in Windows, and fulfills many communications functions. Such software lets you make or take a phone call, manage your phone lists, store an unlimited phone directory, autodial a number by picking it from a list on the screen, and automatically log calls. Some also have answering machine, fax-on-demand, and paging capability.

The CompuNet 2000 from Integrated Technology Inc. (800-393-8889) does it all from your PC keyboard. You replace your keyboard with CompuNet's model, and plug your phone line into the keyboard. You also plug in a handset or headset. When a call comes in, your keyboard rings. *Wow!* The CompuNet 2000 comes with a headset for hands-free dialing. Because it has a separate power supply, you can use the telephone even when the computer is off. And—in addition to the standard 101-key layout—has volume control, mute, redial and flash capability. You can use the hot key function to dial our from any application. All you do is highlight a telephone number on the screen, hit enter and the number is dialed for you.

If you recently purchased a Mac or PC computer, you may already have everything you need to start phoning from your desktop. Many new models are coming prepackaged with computer telephony products including voice mail, answering machine capability, fax-on-demand and data modeming. Phone companies are also getting into the phone software business with products such as Southwestern Bell's PC Phone Manager. Look for packages providing features such as automatic call logging, Caller ID, and Call Waiting.

If you want to go the do-it-yourself route, purchase a voice/fax/data modem for your computer. Most major modem manufacturers offer multifunction modems. Some that come highly recommended are: U.S. Robotics (800-342-5877), Aztech Labs (800-886-8859), Cardinal (800-775-0899) and Connectware (972-907-1093). Connectware's PhoneWorks supports Caller ID, handle hundreds of voice mailboxes, can allow remote notification and access, provides fax-on-demand, a full-duplex speakerphone and very fast data speeds (28.8 Kbps). Installation is reasonably simple—if you're comfortable with a screwdriver, delicate when inserting boards and can tell a dip switch from a diode.

Using your computer as your telephone is not the solution for everyone. You'll have to leave your PC on all the time and, whenever you receive a call, your computer may slow to an irritating crawl. Plus you'll need a goodly amount of hard disk space to store voice messages. Even compressed, a minute of voice can take up to 1.3 megabytes of disk space.

Computer telephony is a fast-changing field. To stay current, get a copy of:

> *Computer Telephony Magazine*
> 1265 Industrial Highway
> Southampton, PA 18966
> 215-355-2866

Stay tuned

Videophones are becoming interesting again, especially those designed for remote conferencing, using whiteboard applications and document sharing. Casio (310-618-9910) makes a videophone that transmits color images and voice over a single analog line. You can even display four images simultaneously or two images via the picture-in-picture function.

Mitsubishi is designing a phone that is the size of two hinged credit cards and weighs less than two ounces. The phone's antenna is a printed circuit board inside the phone. What's next? Star Trek's Tricorder?

Resources

Product comparisons

*Consumer Report*s publishes a survey of both cordless and corded telephones every few years. Check your library or, if you have access to CompuServe, go to the Consumer Reports section (Go CSR) and look under Electronics.

Books

Internet Telephony for Dummies by Daniel D. Briere, Patrick J. Hurley, IDG Books, 1996

6...Phone Systems

• •

At some point, your business may outgrow the stand-alone phones you've been using. Once your needs exceed the capabilities of two- or three-line phone sets and a bunch of extensions, you'll need an integrated phone system. This usually happens around the six-user mark but could happen sooner, especially if you expect to grow fast.

With a phone system, you can maximize your line usage, transfer calls and have intercom capability between extensions. You may save money on monthly line service costs by taking advantage of line sharing capabilities and thus get along with fewer lines. Four lines can easily support eight stations in a standard organization (one that does not have telephone-intensive applications). You may also save money through reducing outgoing call costs by using toll call restriction and least-cost routing.

Phone systems come in six broad categories:

• Key system
• KSU-less system
• PBX
• Hybrid system
• Centrex
• PC-based system

Key system

Key systems were first introduced in the 1920s and were the marvel of the modern office. Since then, they've become electronic, smaller in size, and easier to operate. Key systems are still very common in offices today, though they are losing ground to hybrid systems.

You can spot a key system by the multiple lighted buttons that allow each user to select an outgoing line or answer an incoming one. Instead of having six phones on your desk, each with one line, you can install a key system with six buttons that let you select a line. Having a key system is like having a miniature phone switch in your workplace. The keys (switches) let you share lines with others in your facility, thus reducing the number of lines needed.

Key Phone System

The key system allows your phone system to carry additional signals that reveal line status. These signals might tell the system to light up line lights on some phones, or lock out a particular line. Key systems are good for three to 100 users, depending on the individual system, and come with standard features such as hold and conferencing and optional features such as loudspeaker paging, intercom, call restriction and music-on-hold. Costs vary, depending on the features selected and the cost of installation, but range between $175 to $500 per phone.

Modern key phone systems have several components:

- **Key Service Unit (KSU).** This is the box that houses the power supply, switches and central processing unit that runs the system, and a bunch of cards that control features.

- **Telephones.** Key-set telephones are proprietary and not interchangeable.

- **Add-on features.** These are often circuit boards or electronic units that provide special features such as intercom, music-on-hold or paging.

You cannot use your current telephones in a key system. Nor can you add off-the-shelf telephone hardware (such as a cordless phone) in the future. Adding a modem or answering machine will usually require an expensive adapter or a block.

If you have series wiring, you'll have to rewire because key systems require star (parallel) wiring. See Chapter 2 (Connections) for more information.

Key systems do not require a centralized attendant or receptionist to field calls. Any extension user can answer an incoming call by pushing the corresponding button. Therefore, they're best suited for smaller offices where anyone who answers the phone can take a message or track down the called party.

Focus on a key user

Tom Fox operates the FoxBerry Group, a computer consulting service, based in Monroe, Michigan. He's been using a Starplus 616 from Vodavi for about four years now and "loves it." He and his staff of five currently use seven of the 16 stations available. He has four voice lines; the other three lines are dedicated for data purposes—a modem line, a fax line and a fax-on-demand line.

A favorite feature is the music-on-hold device that plays a customized advertising message when someone is waiting on the line. According to Tom, "Many of our customers comment positively on it." He also likes the visual nature of a key system. The lights let you know at a glance what lines are available and if a call is on hold.

The Starplus system has single line compatibility so Tom was able to attach a modem and fax machine with no difficulty. If his company ultimately outgrows current capacity, Tom can use the same station sets with larger Starplus systems. The starter Starplus system cost him $1,500 (controller box and five phones) or $300 per station. Parts of the system were pre-owned, which allowed FoxBerry to keep costs down.

Tom chose a key system because "a PBX was really more than we needed." The only disadvantage he sees is that the system has no voice mail port, which means that it won't be easy to implement voice mail.

KSU-less system

This is an introductory phone system suitable for 2 to 4 lines and 4 to 8 extensions. Instead of having a phone cabinet (KSU), the smart electronics are built into each individual phone set. These systems are relatively inexpensive, but not as flexible as key systems. Because all of the controls are built into the phone itself, you cannot use your current telephones in a KSU-less system.

KSU-less systems are easy to install. You simply plug the new phones into your existing phone jacks (RJ-11 or RJ-14), using standard two-pair wire and plug the electrical jack into an outlet. You can purchase a KSU-less system at your local electronics store or via mail order and get it running in less than an hour. These systems work with series or star wiring—there is no need to rewire. Costs per telephone set range anywhere from $100 to $300.

Southwestern Bell makes both a three- and a four-line KSU-less phone system. Each system comes with a built-in intercom for handsfree conferencing, line status lights, one-button speed dialing, hold, transfer, three-way calling and a speaker-phone. An LCD display shows the number dialed, duration of the call, and the time and date. The cost per phone is about $250 for the three-line system and $300 for the four-line system. If you installed a three-line system with five extensions, it would run you around $1,250.

The system is remarkably easy to install. For example, to get the three-line system up and running, you plug one line cord into a RJ-14 (two-line) jack to access lines 1 and 2 and another line cord into a RJ-11 (one-line) jack for the third line. You also plug the phone into a wall outlet for electrical power. That's it!

Focus on a KSU-less user
Charlie Swanson runs Edgewater Productions, a film, video and multimedia production house in San Francisco. He recently upgraded his two-line phones to four-line KSU-less sets made by GE. These phones provide two-way intercom, speakerphone, three-party conferencing, do not disturb, line selection and status indicators. The phones required no special wiring. They just plugged into the power and phone jacks like regular single-line phones. Each phone cost about $150 so the total investment was just $1,200. Swanson picked the GE four-line phones because they were "cheaper than a phone system, easier to install, and don't need any special support."

PBX (Private Branch Exchange)

It used to be that you had to have at least 50 users before a PBX was cost-justified. But that distinction is blurring as more manufacturers come out with mini-PBXs at competitive prices. You can usually spot a PBX because you have to dial "9" before making an outgoing call. The "9" takes the place of the plug system operator in old-time PBXs. You may see systems called PABX (the "A" stands for "Automatic") or EPABX (the "E" means "Electronic").

A PBX is more flexible than a key system. You can integrate phone, fax and modem lines and tailor stations (telephone extensions) with features like paging, voice mail, and call pickup. You can use regular, off-the-shelf single line phone sets. You can also install proprietary feature phones sold by the PBX vendor.

Perhaps the best reason for getting a PBX is the control it gives you over phone costs, providing reports on who calls where and for how long. It lets you smoothly integrate phone lines and add or delete features for each phone extension.

PBXs excel at saving you money on your long-distance charges through least-cost routing. This is a feature that uses algorithms to search for the most economical line for an out-going call. You or your vendor have to program the least cost rules and update them as prices change. If your organization is a heavy long-distance user, and subscribes to several long distance services, a PBX could save you up to 20% on your monthly long-distance bills. PBXs require star (parallel) wir-ing. Costs per station for smaller systems amount to around $300-$450, installed.

Using a PBX will also allow you to improve your trunk (phone line)-to-station ratio. Larger PBXs can often get along with one trunk for each 10 users. Unfortunately for small busi-nesses, those economies of scale don't work in the reverse. Smaller PBXs often have a trunk-to-station ratio of about 1:3, which is still pretty good.

PBX users compete for a limited number of lines. If all lines are in use, the next outgoing call is blocked and incoming calls receive a busy signal. Because of this call-blocking, you don't have to pay for excess phone lines that you seldom use. The tradeoff, of course, is that sometimes your customers will receive a busy signal during busy times of the day and your staff may have to wait for an outgoing line.

To avoid the need for a full-time receptionist, you can pro-gram a mini-PBX to ring certain phones in sequence (similar to a hunt group). For example, you might have the phone ring twice at a secretary's station and, if no one answers, ring a couple of times more at the office manager's phone. Then, if still no answer, the phone in your office rings. You'll need a receptionist to handle the attendant console required for larger PBXs.

Hybrid system

Hybrid phone systems are a combination of key and PBX technology. Unlike regular key systems, hybrids let you install your current single line phones on some extensions. Another nice feature is that hybrids have non-squared design. This means that you don't have to have every phone line appear on every phone set, unlike key systems that must have each phone line on each phone. This gives you more flexibility. Some hybrid systems require an attendant; others don't. Hybrids range in price from $300 to $550 per station and require star (parallel) wiring.

Lucent Technology's Partner Communications System is a hybrid system. The Partner is a modular system, which works well for small businesses with plans to grow. Although you may start with only one or two employees, the Partner lets you grow gracefully. The initial setup provides a two-line, six extension system. When business warrants, you can purchase an expansion unit, which will allow you to accept up to four lines and 12 extensions. By adding more modules, the system can expand up to 16 lines and 48 extensions. To install the Partner, you just mount a small box on the wall, plug your lines and phones into the box and you're in business.

The Partner can connect standard off-the-shelf telephone equipment such as single line phones, fax, modem, cordless phone or answering device. It comes with a bunch of built-in and programmable features including privacy, recall, last number redial, intercom, handsfree announce, intercom page, toll restriction, transfer, conference and, of course, hold.

Centrex

Centrex is a service that you can lease, on a month-to-month basis, from your local telco. Centrex gives you phone-system control without having to buy any special equipment. You can select from a broad range of PBX-like services such as conference calling, intercom, call forwarding, flexible route selection, automatic call back, caller ID, call pickup and call transfer.

The features you select are up to you. You can even decide which station sets get which features. For example, you might give managers do not disturb and hotline ability, and restrict warehouse phones to local dialing only.

Centrex is the most flexible of your phone system choices. Because Centrex can accommodate anything from two lines to 20,000 lines (or even more), your growth potential is virtually unlimited. You pay only for the lines and features you need, and can add or remove features, expand or downsize easily. Centrex works with your current equipment. You don't have to purchase proprietary phone sets or adapters to hook up single-line phones.

Because calls come in directly, you don't need an attendant, assuming that all your callers know the direct number they need to dial. If you need some kind of directory or receptionist, you can add on an attendant console or set up an auto attendant system. If you want to add stuff to your Centrex system, work with your local telco to be sure to get compatible equipment.

Centrex is perhaps the best choice if you have more than one location. This is because you can connect multiple sites and provide them with the same features and capabilities as those on your premises. Then you can dial your warehouse down the street, for example, without having to place an outside call, thus saving the per-call charge. You just key in the last four or last five digits of the phone number. All locations have to share the same central office for this to work.

Focus on a Centrex user

Precision Navigation in Mountain View, California, manufactures digital automobile compasses. It started with five people and two phone lines but in less than a year expanded to 30 employees. According to office manager Christine Sherer, "Our Centrex system is currently at seventeen lines and can grow right along with us."

When the company moved its manufacturing operation down the street, they could still keep that group in the same Centrex system. Sherer likes the convenience of being able to dial only four digits to reach them, and "It's not an outside call." This saves them money on each call they make to manufacturing.

Because you don't have any switching equipment on your site, you don't need to worry about housing, maintenance, repair, software upgrades, or obsolescence. Centrex is housed in the telco's central office and tended by 24-hour on-site personnel.

Centrex works with your current wiring so you don't need to rewire. This was a "godsend" when Father Paul Martin of Mission San Juan Capistrano needed a new phone system to connect convent, school and parish. He was told by a phone system vendor that the installers would have to dig up the grounds in order to run new cable. Unwilling to allow this to happen on the historic mission grounds, Father Martin opted for a Centrex system that was installed without digging a single hole.

If you live in an area that experiences power outages or is disaster-prone (such as earthquake-country), Centrex has a decided advantage over premise-based phone systems. That advantage is its power backup system. Centrex service is rarely interrupted by a commercial power failure because most central offices are built to withstand just about anything Mother Nature can hurl at them.

On the down side, Centrex will probably cost you more than a PBX in the long run. Industry experts estimate that a PBX will start to save you money over Centrex after about five years. Then again, considering the speed of telco technological development, that PBX may be desperately out of date by that time.

If you like to control "moves and changes" (moving phones and lines whenever you wish), Centrex will slow you down. You usually have to call your phone company to reprogram before you can make the change. You can sign up for an optional control package that lets you do all this, but it comes with a price.

Centrex prices vary considerably, depending on who provides your local phone service. In Pacific Bell territory, Centrex starts at $15.65 a month per line, plus installation costs of $75 a line and a start-up fee of $200. When comparing costs, be sure to note that Centrex service replaces your current business lines and comes bundled with a group of features that you would have to pay for separately on a business line. The chart below compares a Centrex line with a business line. Each line has Three-way Calling, Call Forwarding Busy, and Call Forwarding Variable.

Charges: Centrex vs. business line

Monthly charge	Centrex line	Business line
Service charge per line	$16.50	$10.32
Three-way calling	included	$3.15*
Call Forwarding-Busy	included	$3.15*
Call Forwarding-Variable	included	$3.15*
Total monthly charges	$16.50	$19.77

* Price computed at 25% discount based on number of features selected. Base price for one feature is $4.20. Centrex prices quoted by Pacific Bell, 1997.

Selecting a phone system
Here are some issues to consider when thinking about a phone system:

Capacity
According to the researchers at Telecom Library Inc., "Buying too small is the biggest and most expensive mistake most buyers make." (Source: *Which Phone System Should I Buy? The Guide to PBXs and Key Systems*, 9th edition, Telecom Library Inc. 1995).

When evaluating sizing needs, you need to know the maximum capacity of the system. Fortunately, this is easy to determine. Almost all key system manufacturers (and many KSU-less and hybrid companies) follow the convention of listing capacity numbers as part the product name. That means that a Lucent Technology Merlin Plus 410 provides for four trunks (phone lines) and 10 stations (extensions) and an Executone 4x8 accommodates four lines and eight stations. A Code-A-Phone 616 can handle six lines and 16 stations.

In case it's not obvious, maximum capacity means just that. If you get a 6x12 system, for example, you can only add 12 goodies on that system. If you need a thirteenth device (phone, fax or whatever), you're out of luck. Time to buy a new system.

If you're thinking of going for a Centrex system, you don't have to worry about capacity because Centrex offers nearly unlimited capacity. You can't outgrow it until you need more than 100,000 lines (don't you wish!).

Determining the number of lines you'll need depends on the type of business you're in and the number of people requiring phone access. There are no universal rules regarding lines-to-people ratios, but many organizations find a 1:3 ratio (one line for each three stations) to be adequate. Here are some examples that might help you sort it out:

* A florist or pharmacist may only need two lines (one for voice orders, and one shared between fax orders and credit card authorizations)

* A real estate office might need one line for every two or three agents (after all, they're out of the office a lot, or should be)

* A travel agency might need two or even three lines for every agent (one for incoming calls, another to use for modeming to online ticket services, a third for outgoing calls)

* A motel may need a lot of stations (one phone per room) but few outgoing lines (maybe one for every twenty rooms)

Years ago, a friend of mine who had recently graduated from med school spent practically all her first month's receipts on (what seemed to me at the time) a whopping big 6x16 PBX. After all, she was a solo practitioner with only a receptionist. Sixteen stations for two people? Sounded crazy. But she was right. Her practice grew and the system grew with her. That same phone system worked for ten years. Only now does she need to replace it with a larger one. Remember, it makes sense to buy more capacity than you currently need.

Suggestions for determining your organization's needs:

✔ Contact your local phone company and ask your business representative for a busy line study. This is a statistical printout of the number and frequency of incoming calls that receive busy signals. Be sure to ask for the cost of the study before ordering one.

✔ Check with your professional association or Chamber of Commerce. They may keep statistics on member's telephone setups and typical costs. Other good information sources are online forums like the Working From Home forum on CompuServe.

Lines to People Ratio

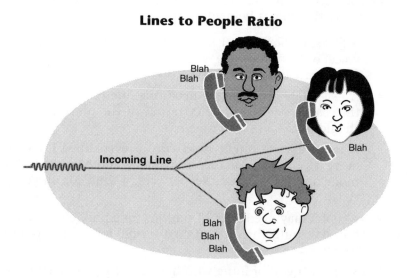

✔ Ask your staff to keep a phone log for a couple of weeks. Divide it into incoming and outgoing. Write down the beginning and ending time of each call. Look for patterns of usage. Do all your customers seem to call right before lunch?

✔ Analyze your current phone use. Do all members of your staff need access to incoming and outgoing lines? Do some people need more than one incoming line at their desk? Does every employee need a phone on his/her desk?

✔ The U.S. Small Business Administration suggests that you allocate a line for each employee that spends at least a third of their day on the phone.

Be sure to allow for extra phone lines to accommodate fax machines, modems, credit card authorization terminals, and answering machines. Some organizations install a phone outside the entrance for safe after-hours entry. You may also need a phone in the reception area, the copy room or back in the warehouse.

Decide whether you want to hardwire certain peripherals (such as credit card or check authorization lines), or run them through your phone system. It probably makes most sense to have dedicated circuits for certain lines, depending on your business. After all, you might not want check-paying or credit-card paying customers to wait while your cashier queues for an outside line.

Compatibility

Phone systems sound like a great idea—until you try to add a modem or a cordless phone made by another manufacturer to your communications system. Key phone system signals might tell the system to light up line lights on other phones, lock out a particular line, play music for a holding caller, or even open a door.

Unfortunately, there is little industry-wide standardization— each key system manufacturer handles signaling a little differently. This means that you probably can't use another telco manufacturer's device in your system and are limited to proprietary equipment. This could be serious and expensive. Just adding a modem may require separate wiring or the purchase of some kind of converter.

This sad situation is beginning to change. Look for systems that advertise Open Application Interface (OAI). The operant word here is *open*. If your system is designed with OAI, you can use your computer to customize your telephone switch, integrate call processing with computer databases and often mix and match equipment from various manufacturers. Standards are still being settled however, so shop with care and ask a lot of questions.

Migration path

Many key system and PBX manufacturers offer modular systems that can grow with you. You begin with a small starter set and can upgrade to larger systems while still using their original hardware, including phone sets and control units. For example, the Lucent Technology Partner phone system can be purchased in modules. You can start with a two-line, six-station system and add modules that allow you to expand to four lines and 12 stations, and ultimately to a 16/48 configuration.

Another interesting option is the ability to combine phone systems. You'll find companies offering combinations of Centrex with a PBX, or a key system running "behind" a PBX. This is sometimes called "piggybacking." Here's how it works: You have a PBX, but want a particular department or group to be more accessible to outside callers. So you install a key system (or Centrex) in that department. Then, rather than calling your main number, and getting the receptionist to steer the caller in the right direction, piggybacked systems allow callers to dial in direct instead of calling the main number.

Digital or analog?

All modern PBXs and hybrid systems, as well as higher-end key systems, are digital. This means that the electronics that run the system use digital rather than analog signaling. Having a digital phone system will give you cleaner sound and more intelligence. For example, your digital phone set can display the station number or perhaps the name of the person calling from within your system. Often, the buttons on a digital set are programmable, allowing you to tailor the system for easy one-button access to often-used features or dialing codes.

Don't confuse digital phone systems with digital lines. Most digital phone systems today are designed to work only with analog lines. To work with digital lines such as ISDN or T-1, you'll need a phone system that is ISDN-BRI (gives you two

64-Kbps lines) or ISDN-PRI-compatible (gives you 23 64-Kbps lines). For more information on ISDN, see Chapter 4.

Cost

How much will a phone system set you back? Depends. Don't you hate that answer? But in the phone system world, you don't often find price lists. And a major portion of the price may be the charge for the installation itself. Also, other variables are the features you select and the type of station sets you choose. If you call for a quick quote, often you'll get nothing more than a vague answer or, at best, a range.

Phone systems are often quoted on a cost per station basis. This is derived simply by adding up all the costs of hardware, installation, warranty and other extraneous charges and dividing it by the number of extensions. Thus a 4x8 Execu-tone key system quoted at $1,799 would cost out at $450 per station. This makes it somewhat easier for you to make cost comparisons.

Although we all want to save money, buying a phone system solely based on price considerations is false economy. Harry Newton, publisher of *Teleconnect* magazine, gives three reasons why it's stupid to shop price exclusively:

1. The people who answer your phone cost a great deal more than your new telephone system. If they hate the system, they'll do a lousy job of answering the phone.
2. Ninety percent of your customer's contacts with your organization take place initially over the telephone. You want that first impression to be as positive as possible.
3. A phone call is much less costly than other types of communications, especially when you factor in employee time.

If you need to save money, consider buying a used or refurbished system. Savings from 30 to 50% are common. You can locate used equipment by calling a local installer or pick one up at a business auction. You might also contact the North American Association of Telecommunications Dealers (NATD, 561-266-9440). They can provide you with the names of NATD-certified secondhand equipment dealers in your area.

Features

Phone systems usually come with loads of features—probably more than you need and maybe more than you want. I've listed some of the most useful below. You won't find all these features in every phone system. The feature set depends on the technology (key, Centrex, etc.) and the price. I have also included some features that are more appropriate for larger organizations because, after all, you're going to grow.

ACD (Automatic Call Distribution)

Lets you program routes for incoming calls so that the calls are distributed to a designated group of stations based on which agent has been idle the longest. If all stations are busy, calls are queued and processed on a first-in/first-out priority. ACDs are more common in larger phone systems. ACDs often provide useful management reports such as which agents were idle the longest, the number of calls abandoned in queue, and the number of callers waiting.

Alarm signals

You can connect alarm systems—smoke detectors, fire alarms, door bells, intruder alerts and the like—to your phone system. Then when an alarm condition is detected, you can be alerted by a ringing phone, a horn over the loudspeaker, etc.

Automated attendant

A program that answers your phone, plays a pre-recorded greeting that prompts the caller to choose a route for his or her call. May have one or many levels that branch out to provide choices. The automated attendant prompts the caller to dial the specific extension number of the party desired, dial "O" for operator, or give helpful messages to the caller.

Automatic call back

Allows office mates to select automatic retry if a dialed station set was busy. Can be used as a message reminder.

Call accounting codes

Gives you the ability to enter a code (often several digits long) to charge a call to a department or project number instead of to the calling station. This is especially useful if you need to bill specific projects or clients for your time. You can usually set up the system to either force the code, force and verify the code (check that the code entered is on your list), or accept an optional code. Forced and verified codes have the additional benefit of curtailing unauthorized access to your telecom system. See *SMDR*, later in this chapter.

Caller ID display

Displays the phone number and, in some cases, the name of the calling party if the information is available. You must sign up with your telephone company to get Caller ID service for this feature to work. Some phones also have memory capability for storing a running list of incoming caller's numbers. This memory feature even records the phone numbers of callers who hung up before you answered.

Call forwarding

Lets you forward your incoming calls to another station. Some systems reroute calls when the intended station is busy or doesn't answer after a specified number of rings. This is similar to phone company Call Forwarding. Of course, you can only forward to other stations within your phone system. The system may have call forwarding as well. System-wide call forwarding allows you to forward all your after-hours calls to an answering service or to voice mail, for example.

Call queuing

Keeps a running queue of all incoming calls and delivers them to call answering staff in the order received. This is a simplified version of automatic call distribution. See *ACD*.

Call park

You place a call in a waiting condition, page a third party, and have that party pick up the parked call from any station in the system. This method is often used by systems that have a limited number of holding paths.

Call pickup

Allows you to program call pickup groups that may retrieve a call directed to another station in the same group. For example, if the phone in the next office is part of your pickup group, you can transfer the call to your phone and answer it, thus saving steps.

Call transfer

Lets you transfer a call to another station. The call will ring and flash (light up a LINE key) only at the station you called. This lets the person transferring the call process calls faster because s(he) doesn't have to wait for a reply from the party being called.

Camp on busy

If you're dialing an extension within your office and the line is busy, you can push the CAMP ON button or dial a code that lets the called party know that there's an internal call waiting. Your phone will ring you back automatically when the called party is off the line. This feature is also called Call Waiting on Intercom.

Conference

Lets you add on other persons to an existing conversation. Useful especially for professionals who often need three-way conversations.

Data privacy

Allows a user to specify that no interrupt tones, line signals or pauses be allowed to interrupt a call. This safeguard eliminates the danger of inadvertently transferring a call to a line being used for a modem during a data transmission.

DID (Direct Inward Dialing)

A feature, used by PBXs and Centrex, that allows the caller to bypass the receptionist. The call is passed straight through to the dialed extension. This, in effect, gives each station user an exclusive telephone number for incoming calls. Saves call handling time.

DISA (Direct Inward System Access)

Provides the ability to designate a line or group of lines that may be accessed from a remote location. Allows the incoming caller to use system features or special circuits. You could, for example, call in to the office from home and dial a code that lets you take advantage of your office long-distance dialing plan discounts, charging the call to your office line.

DISA has a big disadvantage. It is not secure, therefore phone hackers could get into your system and ring up an enormous phone bill ($50,000 over a weekend is not unusual) for which you are responsible. The telephone industry reports that DISA phone fraud costs businesses over $1 billion per year.

Beware: Hackers especially like to prey on small businesses because smaller organizations usually don't have phone fraud detection equipment.

If you still think you need DISA, be sure to create complex hard-to-hack codes (don't use birthdays, anniversaries, your name, dates), change codes often, and always change codes whenever an employee leaves.

Distinctive ringing

Calls ring differently depending on whether they originate internally or externally. Lets you distinguish between an intercom call and a client call, for example.

Do not disturb

If you invoke DO NOT DISTURB mode, your station appears busy to incoming telephone and intercom calls. Some systems give the caller a distinctive tone, alerting them that you are in but not available. This is great if you need to get a project done and can't afford interruptions. It's also useful when you're on the phone and do not want to be disturbed by the intercom.

DSS/BLF stations

Direct station selection/busy lamp field stations are telephone sets that allow visual recognition of busy line status, and let you select a line or station just by pressing a key on the phone.

Exclusive private line

Reserves a line for the exclusive use of one station, often an executive privilege. Allows the user to receive confidential calls that bypass the receptionist. Also ensures the availability of an outgoing line even at the busiest of times. You rarely see this feature in smaller phone systems.

Executive barge-in

Lets you program privileges that allow a certain person or persons the ability to listen in or override privacy features such as Do Not Disturb. Depending on the design of the feature, the employee whose conversation is being overheard may or may not hear a beep tone. This feature is often used by managers to observe worker's telephone performance and can be a useful training tool. It can also be a major intrusion and seriously harm morale. This feature is also called Executive Override.

Fax detection

Automatically routes incoming fax calls to the extension where your fax machine resides. Eliminates the need for a separate fax number.

Flexible line assignment

You can program your phone system so that certain departments or individuals have groups of outside lines assigned exclusively for their use. Use this if you want to ensure that your sales force, for example, has sufficient lines so they don't have to wait for an outgoing line.

Group intercom

Establishes a private intercom line between members of a preselected group. Lets you reach anyone in your group fast without having to dial an extension number.

Handsfree offhook announce

Lets you page and respond to internal paging through the station speaker without having to pick up the handset. Your secretary could push a button and notify you of a call or an onsite visitor. You could respond to your secretary, asking him or her to take a message, for example. Much nicer than just hearing a buzz to alert you to a call or visitor. There's a down side too. Unless you have a privacy button (and use it), anyone can interrupt you at any time or, even worse, listen in to whatever goes on in your office without your knowledge. Sort of reverse bugging.

Hot line

You can program the phone to always dial the same number or extension number whenever you pick up the handset or punch a designated key. Saves time if you frequently call a specific number.

Intercom paths

The number of possible simultaneous internal conversations (intercom paths) will not necessarily be the same as the number of outside lines available. Some systems support only one or two intercom paths. If you use (or plan to use) intercom widely, be sure to check this out.

Intercom

This is an internal communications system that lets you communicate with another phone in your office. This may be done by dialing an extension or pushing an intercom button that buzzes a particular extension. The dialed or buzzed party then picks up the phone to carry on a conversation with you. See also *Paging*.

ISDN compatibility

ISDN technology lets you turn a normal twisted-pair copper phone line into three digital channels (virtual phone lines). If you combine two of the lines, you can use ISDN to provide a broadband channel for bulky information transfer. You may want to add ISDN in the future. To stay flexible, look for a phone system that can work with

ISDN. About half of them do. For more on ISDN, see Chapter 4.

Integrated voice mail

Some phone systems come with a built-in or optional private-labeled voice mail system. It may cost less to get an add-on device than to purchase voice mail as a phone system option.

Least cost routing

Lets you predetermine call routing based on the least expensive routing available at the time the call is placed. You (or your phone system vendor) program choice rules for selecting phone lines for outgoing calls. Organizations using least cost routing schemes report toll charge savings of 20% to 30%. This feature is appropriate for companies that use a combination of dialing options such as multiple carriers, DDD (Direct Distance Dialing), foreign exchange, and tie lines. This feature is also called Automatic Route Selection.

Message waiting

The attendant can activate a lamp on a station set to indicate that information (such as a message) is waiting. This is especially useful in a busy office and reduces the frequency of lost or forgotten messages.

Music/message on hold

You can hook up to a local radio station, play a classical tape, or play advertising or informational messages for callers on hold. Allows the caller to know that he/she is still connected. Some key systems contain a music synthesizer that plays music-like sounds. I don't know about you, but I'd rather listen to pounding heavy metal or country corn—anything but elevator music.

Tip: If you use local radio, be sure to listen to it yourself. Make sure your callers won't be irritated by rude or loud announcers, soppy music or, worst of all, commercials from your competitors. It has happened before.

Night service

Provides the ability for incoming calls to be answered at night when there is no attendant to answer and transfer the call. You can program the night bell to ring at a specific station (security guard), or at a specific department that regularly works late. Most often found in PBX systems and some hybrids. See *TAFAS*.

OPX (Off Premises Extension)

Allows you to connect a station that is physically located in another building to the office telephone system. The OPX may be a person's home or another building.

Paging

This has nothing to do with dialing a pocket pager or beeper. Paging, in the phone system world, refers to announcement features such as:

Internal paging - lets you make simultaneous announcements to all stations by using an internal speaker located in each station set.

External paging - lets you make amplified announcements over speakers or paging horns. Most useful in warehouses, construction yards or large open spaces where a telephone or internal paging would not be practical. Remember those speakers when you're shopping for a used car? "Frank, you have a phone call on line four."

Zone paging - may be internal or external. Lets you make paging announcements to specific areas within the office or facility.

PC programming interface

This allows you to program your phone system using software running on a personal computer. Simplifies maintenance and updates.

Peripheral adjunct

Lets you connect an answering machine, fax machine, modem, security system, credit verification or other POS

(Point Of Sale) terminal, cordless phone, or headset to the phone system or a station set using an adapter.

Pooled access

Lets you share outgoing lines without the need for expensive telephone sets with direct line appearance (a light for every outgoing line). To dial out, you just press a group access key or dial a specific code.

Power failure protection

Phone systems need electrical power in order to operate. Unlike ordinary phones, which get their ringing power from the telephone line itself, telephone systems get their ringing power from an electrical outlet. That means that if there's a power outage, your phones won't ring. Since this situation is unacceptable to most businesses, telephone system designers have installed backup battery packs or a bypass system which allow at least a few of your phones to ring if power is out.

Recall from hold

If a call has been left on hold for a specified time, the phone will ring or flash as if a new call was arriving to remind you of a holding caller.

RS-232 jack

Some digital telephone station sets come with a RS-232 jack that can be connected to a like RS-232 port on a personal computer, fax, modem, or printer. This allows you to have a computer terminal at the same location as a telephone set without having a separate dedicated data line. This feature is also called Dual Port Design.

Single line telephones allowed

Allows you to attach non-proprietary off-the-shelf telephones to the telephone system.

SMDR (Station Message Detail Recording)

Provides you with a chronological record of all outgoing calls. Information generally includes station ID number, date and time of call, duration of call, number called, and trunk (telephone line) used. You can use this report to bill calls back to departments or tenants, check for unauthorized call use, and spot lengthy calls. Frequent reviews of SMDR reports can significantly reduce your local and long distance bills.

Speed dial/redial

This works just like speed dial/redial on a stand-alone phone. You program frequently called numbers and dial them with a one- two- or three-button code. Redial (also called Last Number Redial) automatically dials the last number you dialed.

Station-to-station messaging

Allows members of your organization to create and send brief alphanumeric messages that will appear on the telephone set's display or illuminate a light indicating that a message is waiting. Convenient for informing a coworker that you're OTL (out to lunch) of GFD (gone for the day.) You're not limited to these abbreviations. You can spell out anything you want.

TAFAS (Trunk Answer From Any Station)

A type of night service where incoming calls activate a special ring or alerting signal (bell or gong). To answer the call, you pick up the handset and dial a special code. This feature is also called Night Answer Mode.

Tenant service

A feature of certain phone systems that allows you to create system subsets. Each tenant can have its own attendant console and, possibly, its own telephone lines. Useful in office buildings that are shared by more than one tenant. Reduces the cost of communication services.

Toll restriction

Lets you program stations to restrict their access to long distance service or specified toll calls. You can restrict access to one or more area codes or even specific area code and prefix combinations. Saves money on your long-distance bill.

Trunk queuing

Enables members of your organization to "get in line" for an outgoing trunk (phone line). When an outside line becomes free, the system rings the next station user in the queue.

Trunk-to-trunk transfer

Lets you transfer an incoming call to a number outside your phone system such as a voice mail service.

UCD (Uniform Call Distribution)

Similar to ACD except that UCD permits a group of incoming lines to be answered by a group of stations using a round-robin or top-to-bottom distribution scheme rather than "most idle" schemes.

Voice mail

A built-in capability to set up and maintain multiple voice mailboxes.

PC phone system

The marriage of computers and telephone systems is just beginning. Look for more and more examples of computer telephony designed for the SOHO market in the coming months.

The Dash Open Phone System combines personal computer technology and telecommunications to create an advanced phone system for small and medium businesses. It's a PC-based expandable PBX with built-in capabilities such as auto attendant, call queuing, voice mail, call transfer, handsfree intercom and "meet-me conferencing." The smallest size supports up to eight lines. By adding expansion modules, the system can grow to 128 ports (half of them could be incoming lines) without having to replace the central processor.

Optional features include Caller ID, music on hold, security control, and ACD. Another useful feature is the ability to connect an off-the-shelf single line phone to your PC. This is done through a Windows-based telephony manager, that allows a standard phone to juggle as many as four calls at once. Features include conferencing, call transfer, message waiting, and voice mail. To dial out, you just call up your phone directory, point and click with your mouse. (Dash Inc., 800-464-DASH).

Which way to go?

Deciding what type of system to get is not easy. Here's a checklist to assist you in your search:

✔ Project your estimated growth for the next five years. Then double it. You need to get a system that can gracefully grow to that size. Plan on a five- to eight-year lifespan.

✔ What do you pay for current monthly phone service? Will a phone system allow you to cut back on the number of lines coming in? Or reduce the number of phone company features (like three-way calling) that you may be paying for? If so, how much will that save you now? Over the next five years?

✔ Read the features section in this chapter and then make three lists: must have, like to have, nice to have. Can you get those same features in stand-alone phones?

✔ Can you use your current equipment (phones, accessories) with this system, or do you have to buy all new stuff—and only from the vendor?

✔ Will you have to rewire? If so, figure that cost in.

✔ Is your business cyclical? Campaign organizations and seasonal businesses aren't the only ones that need to downsize as well as grow rapidly. If you have major ups and downs, you don't want excessive hardware, Centrex may make the most sense.

✔ Are capital funds limited or difficult to spend? Many schools, governmental agencies and charitable groups opt for Centrex service because it's easier to fund.

✔ Watch out for long feature lists. Just because a phone comes with "57 labor-saving features" doesn't guarantee that you or your staff can use them. How often have you heard "if I lose you" as the preface to transferring a call? What does that tell you?

✔ Can you postpone the phone system decision for a few months? If you can, I'd recommend that you do. Telephone systems are changing rapidly, prices are coming down, proprietary systems are losing ground to PC-based telephony and open systems. Because standards aren't completely firm, it may be prudent to wait.

One final note: While doing research for this book, I talked with lots of businesspeople. Some were content with their phone system. Many were not. Pick a system that can grow with you. Otherwise, you'll end up with an expensive mistake. Consider this comment from Pacifica real estate broker John Doyle, who recently purchased a system. He warns, "As soon as you buy it, it's out of date."

Resources

Books & magazines
Which Phone System Should I Buy? The Guide to PBXs and Key Systems, 9th edition
Telecom Library, Inc., 1995
800-LIBRARY
A useful guidebook that lists more than 180 different phone systems (PBXs, key systems and hybrids). It also includes features, capacity, specs and ballpark prices.

Computer Telephony: Automating Home Offices and Small Business, by Ed Tittel and Dawn Rader, Ap Professional

Teleconnect magazine
Published monthly by Telecom Library, Inc.
800-677-3435

Computer Telephony Magazine
1265 Industrial Highway
Southampton, PA 18966
215-355-2866

7...Mobile Phones

Does your business take you out of the office a lot? Do you need to keep in touch with your customers at least as often as they need to call you? Are you always on call, or do you spend a lot of time in your car or at a field site where a phone is not readily available? If so, think wireless. With a wireless phone, you can conduct business almost anywhere—restaurants, ballparks, parking lots, grocery stores, building sites and roadways—to name a few. Wireless phone service can add extra hours to your working day and provide a ready link to emergency services. By the end of 1996, there were more than 43 million wireless users in the United States, accounting to about 16% of the US population.

This chapter is about wireless phones—including portable, transportable and car phones. I've added advice on selecting a cellular phone and carrier and information to help you keep your cellular bill under control. I've included information to help you decide whether analog, digital or PCS (Personal Communications Services) is best for you.

Note: Cordless phones, which rely on a different technology than mobile phones, are covered in Chapter 6.

A brief history of mobile services

Cellular service is licensed by the Federal Communications Commission (FCC). The FCC establishes how many channels are available, and it licenses the cellular operators (called *carriers*). Until recently, the FCC divided the country into 305 metropolitan areas and 428 rural service areas and allowed no more than two cellular carriers to compete in each area. So, your choice of carrier was quite limited. This situation is beginning to change.

In 1996, the FCC auctioned off a group of licenses to allow communications companies to sell a new generation of mobile phone service. Called Personal Communications Services (PCS), this new technology allows for expanded services, better sound quality, and lower prices. A PCS phone operates on a different frequency than a cellphone, and it can operate with less power. This results in longer battery life and lighter phones. Another expected advantage of PCS technology will be greater competition. Instead of the two cellular providers currently offered in an area, there could be as many as eight PCS-service providers in a given market.

That's the good news. The bad news is that, until a clear winner appears in the fight for wireless buyer loyalty, there will be four incompatible transmission standards. Buy the wrong one, and you may be out of luck. To make a decision about which mobile technology to go for, you're going to need to know how to decipher wireless acronyms and plow through competing claims. This chapter will help you through it.

How cellular works

Cellular phones are basically two-way radios that operate on certain air wave frequencies. Unlike typical two-way radio communication, however, a cellular system can connect you to the wired telephone network. Also, cellular systems, unlike two-way systems, are full duplex. This means that both parties to a conversation can talk and listen simultaneously, just you would on a normal telephone line.

A cellular system is made up of hundreds of miniature radio communication transmitters and receivers, each situated a few miles from the adjacent station. Each of these stations is called a cell. As a cellular user moves through the service area, the call is electronically switched from one cell station to the next, allowing calls to continue without interruption.

How Cellular Phones Work

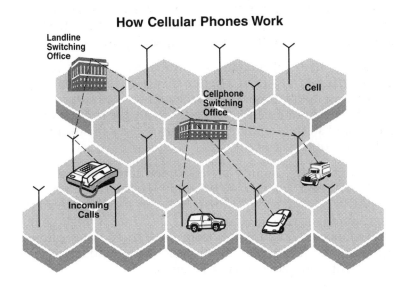

None of this is apparent to the user. As far as you are concerned, you place a call in a manner similar to using a land phone, except there's no dial tone and you press the SEND key when you're done keying in the number. And when you want to hang up the phone, you press the END key. When your phone is off, you can't make or receive calls. People who call you will hear a message indicating that your phone is not available. If you want to receive calls, you place the phone in standby mode. Then, if a call rings in, you press the SEND key to answer.

Cellular isn't perfect. Because radio waves bounce off walls and are diminished by trees and other greenery, they can get distorted. You may get disconnected, have calls fade in and out, or even hear crosstalk (hearing another cellular conversation on your frequency or channel in celltalk). And you'll experience great difficulty making or receiving calls in tunnels or canyons.

Digital cellular

Not all metropolitan service areas offer digital service today, however, digital is the future of cellular service. Digital cellular phones are more expensive but offer enhanced features. These include call scrambling for privacy, selective call acceptance, priority access, call forwarding, voice mail message notification, alphanumeric messaging, and Caller ID.

Digital has many advantages. Because more channels can be derived, digital calls stand a better chance of ringing through during peak calling times (such as 9 a.m. on a Monday morning). Signal sound quality is superior to analog because interference and static are filtered out. This reduces the likelihood of dropped or blocked calls.

Digital airtime is currently priced somewhat lower than analog. The reason for this is twofold: (1) Digital is actually cheaper to operate because the carrier can increase the number of simultaneous conversations derived from a single channel (estimates show increases from three to 10 times current capacity), and (2) carriers are pushing digital and are, therefore, offering attractive pricing schemes.

Competing technologies

Unfortunately, cellular phones come in different incompatible flavors. In many cases, your choice of phone will lock you into a particular service provider. There are several technologies, each with its own transmission standards. These are the choices available:

- **AMPS**
  Advanced Mobile Phone Service. This is the standard 800-MHz cellular band where all analog cellular systems operate. AMPS is widely available.

- **CDMA**
  Code Division Multiple Access. A type of digital phone. Also shares the same 800-MHz cellular band as analog cellular.

- **TDMA**
  Time Division Multiple Access. This type of digital phone also operates in the same 800-MHz cellular band as analog

cellular. TDMA service is more widely available than CDMA.

- **PCS**
Personal Communications Services. These digital phones operate at 1900 MHz and use one of three different transmission standards - TDMA, CDMA and GSM.

- **GSM** (also called PCS-1900)
Global System for Mobile Communications. This is the digital standard throughout Europe, though on a different frequency. This means that your American model GSM phone probably won't work in Europe.

Bottom line

Analog user: Analog cell phones all use the same technology, so you can use your phone wherever you are—as long as you have roaming capabilities.

Digital phone user: These standards won't be such a headache for you if you use your wireless phone only in your own service area. But, if you plan to travel with your phone, you'll need to know what works where. Alternatively, leave your phone at home and rent one at your destination.

Is any one format better than another? Not that the normal user would notice. Industry insiders note that GSM has been out longer so has undergone more testing than the others. Some phones are equipped to handle multiple technologies.

Cellular charges

In addition to a one-time set-up fee, you will face three types of monthly charges: a monthly subscription cost, a charge for landline services provided (calls for directory assistance, for example) and an airtime charge (the actual time you spent using your cellular phone). The first two charges are fixed; it's the airtime charge that can get you.

Unlike land-based phone service, the cellular subscriber pays for both incoming and outgoing calls. In addition to your monthly access charge, you pay per minute charges for the time you spend on the system. The cost of airtime varies considerably depending on the time of day. Cellular carriers usually define peak time as Monday-Friday, 7 a.m. to 7 p.m. You can usually reduce your charges by shopping for promotions, discounts and packaged regional service plans.

The best way to cut your costs is to give out your cellular number only to your best clients, or to combine cellular with a numeric paging service. Arrange for your voice mail or paging service to beep your pager and display the number of the caller. Then you get to decide—dial out now on your cellular phone or later from a less expensive landline. That's what Dave Steele, a maintenance management consultant in Mill Creek, Washington does. Steele explains that using a pager saves him $20 to $30 a month in cellular charges. He calls his voice mail service to check on messages only when the pager alerts him. "Working from a cellular phone," he says, "I hate to spend the 45¢ to 90¢ a minute just to find out that I have no new messages."

Because the carrier charges for calls placed by you the moment they are connected to the cell site, you may get charged for calls that never went through. Long-distance adds up fast because you pay two charges: airtime and normal long-distance charges.

Carriers offer a variety of pricing plans that cater to light, medium and heavy phone users. The lowest monthly rates are often called "Night Owl," or "Emergency" plans. These give you a cheap rate for after-hours calls, but calls during business hours are billed at an astronomical rate—up to 90¢ per minute.

Bottom line: Although the average monthly cellular phone bill in 1996 was $47.70, plan on spending quite a bit more if you use the phone as a business tool. A $100 monthly plan with an average use of 250 minutes a month is typical. That adds up to about $1200 a year.

Roaming - using your phone outside your service area

When you sign up for a cellular service, you're allowed to make and receive calls within your calling area. If you travel to another city, and want to use your cellular phone, you'll be operating in another calling area and will be *roaming*. Your calls will be charged at the rate of the host carrier, and they could end up being quite a bit more expensive that your budget allows. Some carriers charge a roaming charge that can run as high as $3 a day and $1 a minute. Fortunately, the trend is toward elimination of roaming fees.

If you plan to roam a lot, consider installing more than one NAM (Numeric Assignment Module). A NAM is a chip installed by the cellular dealer that contains your unique cellular phone number. Most cellphones come with room for two NAMs; some allow up to six. If you have multiple NAMs, you can register your car phone with several different cellular providers.

Another option is to choose a cellular provider that offers same-as-home rates. This means that the provider has a regional network and charges the same rates within the region as those in your home area.

Roaming costs add up even faster when you receive incoming calls. That's because you are charged long-distance forwarding charges for the connection between your home cellular service and the current location. For example, you sign up for cellular service near your office in Dayton, Ohio and are on a business trip in New York City. You've set your cellular phone on roam. One of your clients, Anne, is supposed to meet for dinner but calls you from Grand Central Station to tell you she is running late. Anne will have to pay for the long distance charges for forwarding her call from New York to Dayton. You will be charged for the call-forwarding costs from Dayton back to New York. Worse yet, the calls are often charged at daytime rates, regardless of the time of day.

One way to avoid long-distance surprises is by using your calling card for outgoing calls. For incoming calls, your best bet is to leave your roaming access number with people who may need to reach you in an emergency. Alternatively, sign up for caller notification service with your cellular carrier. This service plays a message to your callers and gives them the roamer access number.

A useful source of information for roamers is a cellular directory containing maps of cellular calling areas and addresses and telephone numbers for cellular carriers. See *Resources*, at the end of this chapter, for suggestions.

Privacy
Although cellular calls are not as leaky as cordless phone calls, electronic eavesdropping is still possible. Conversations can be overhead by some types of radio receivers and scanners. If you are unlucky (or famous, like Princess Di), your conversation may be overheard. So, if your business requires that you make calls that are sensitive and must be secure, avoid cellular phones.

For the same reason, you might not want to give out your credit card numbers over cellphones. You could buy a scrambler, but the person on the other end would need a matching descrambler. Or you could buy a high-end phone with voice encryption. Digital cellphones are much more difficult to tap than analog cellular phones (more on this later).

Cellular phone features
Here is a sampling of some of the more common features available:

A/B switching
Provides easy switching between carriers. This is particularly useful when roaming. You have to have a dual-NAM phone to use this switch.

Alphanumeric memory
This lets you type in names to go along with the numbers you have stored in memory. This isn't as simple as it sounds. Since there are only eight buttons on your phone with letters on them, you have to follow a weird coding scheme (*Example:* for the letter C, press the 2 key 3 times; for the letter U, press the 8 key twice).

Answering machine-like capabilities
Look for more and more cellular phones to have the ability to take messages (some with built-in time stamps).

Any key answer

Lets you answer the phone by hitting any key. Especially helpful if you're driving and your pocket phone rings.

Automatic credit card dialing

Gives quick access to your choice of long-distance carrier. If you make a lot of long-distance calls over your cellular phone, this feature can save you money.

Automatic NAM selection

Your phone will detect which call area you are in and set the correct NAM for you, ensuring that you won't incur roaming charges. This also saves you time because you don't have to figure out the NAM you need.

Battery level meter

A graphic display of battery strength. Try to get a phone with one of these gauges; it will let you know how much power you've got left. Earlier and cheaper models have a low-battery light.

Battery life

This is usually listed as two figures: talk (air) time and standby time (waiting mode). Low-end phones have about 60 minutes of talk time; high-ends go up to 2.5 hours. Standby time ranges are from 12 to 72 hours.

Built-in pager

This useful feature saves you from having to carry two devices.

Call timer

Many phones have more than one call timer option such as call-in-progress timer, resettable cumulative timer and last-call timer. A handy feature is a one-minute alarm which beeps you but not the caller.

Data port

If you want to send a fax or download a file remotely, you'll want to link your cellular phone with a notebook computer or portable fax. For more information, see *Cellular Fax*, later in this chapter.

DTMF

This stands for Dual-Tone Multi-Frequency, a fancy word for touch-tone. If you plan to use your cellphone to call your bank for your balance, transfer funds, call into voice mail, or effect other automatic call-processing, you'll need to be sure that your cellular phone sends valid touch-tone signals. To be sure, get a phone with "extended DTMF."

Dual mode

This allows you to switch your phone between analog and digital technologies. This will become increasingly useful as more local carriers add digital service.

DualNAM

This will help you save on roaming charges because you can register for home service in two cities.

Electronic volume controls

Separate settings for earpiece, ringer and keypad volume.

Lock

Disable transmit and receive functions to prevent unauthorized use of your phone.

Memory

You'll want your cellular phone to have memory dialing. How many numbers do you want to store? 30 is standard; many phones can handle 100 or more. For those rare human beings that can remember more than 100 phone codes, a few phones store up to 300.

Missed call counter

Registers unanswered calls received while the phone is in standby mode. Especially helpful if you have voice mail.

Multiple NAM capability

A NAM is a chip installed by the cellular dealer that contains your unique cellular phone number. If you have multiple NAMs, you can register your phone with several different cellular providers. Many advanced phones offer from anywhere from three to eight NAMs.

Numeric answering mode

This setting answers incoming calls and allows the caller to enter a call-back number which then waits in temporary storage until you pick it up.

PCS compatibility

A few high-end analog cellphones will operate with PCS at 1900 MHz as well as at analog AMPS 800 MHz. Sometimes called dual-band.

Remote control

Lets you call your cellular phone from another phone to power down or lock up the device.

Ring tone selection

When you're in a crowd, how will you know whose cellular phone is ringing? One way is to have a ringer selection feature that allows you to select the ring of your choice. Even better is no ring at all; you can now get a phone that vibrates.

Scratchpad

Allows you to type a number into temporary memory while on a call. Silent scratchpads are best.

Signal level meter

Displays the strength of your cellular signal. Can be useful in notifying you that you're moving into an area with a weak signal.

Voice message waiting indicator

Handy if you're out in the field and get frequent voice messages. Saves having to call in to your service.

Types of cellular phones

There are three basic types of cellular phones:

- Portable (pocket) phone
- Mobile (car) phone
- Transportable phone

Portable phone

The smallest type of cellular phones are portables (often called pocket phones) that run on rechargeable batteries. The heavier the battery, the longer the time between charges and, therefore, the longer talk-time possible. Portables weigh anywhere from three to 18 ounces. Price is inversely related to weight: the lighter the phone, the heavier the cost.

The maximum power level available for portables is 0.6 watts. This is plenty for normal urban or suburban use. But, if you often travel to rural areas, suburban fringes, or concrete canyons, you might need the greater power available with transportables and mobile phones. Of course, if you want to take your phone everywhere you go, you'll probably want a portable.

Battery basics

Batteries are a big deal with portables, so the choices are wide. Some models let you recharge your battery while you're on a call; others don't. Some portable models come with an extra battery. If not standard, buy an extra battery anyway. Otherwise, you'll kick yourself when you run out of power during that all-important deal-maker call.

Battery chargers range from small portable trickle chargers to desktop-sized rapid chargers. If your model uses nickel-cadmium batteries, get a charger that has a discharge button so you can fully cycle the battery. Otherwise, your battery life will get shorter every time you charge.

This occurs because battery memory build-up decreases the cell capacity of the battery when batteries are only partially discharged and then recharged. New on the market are nickel metal hydride (NIMH) and lithium-ion (LISB) batteries that are supposed to resist memory loss. Lithium-ion batteries are heavy-metal free, a bonus for the environment.

If you're having trouble finding a replacement battery, or your cellular model is no longer being produced, contact 800-Batteries, a discount mail-order company that specializes in batteries for notebooks, laptops, cellphones and the like. If they can't locate a battery that works with your model, they'll actually rebuild your battery for you. Call them for a free catalog (800-BATTERIES).

Car phone

Mobile or car phones are permanently installed in a vehicle. They are powered by the car's battery so you never have to think about battery life or where you can recharge your phone. Car phones are known to have the best reception of the three types of phones. This is largely because the car itself helps serve as part of the antenna, and they operate with more power.

Car phones are the cheapest cellphones available. They're economical because there are no power supply problems and there's no need to minimize weight. Car phones are basically maintenance-free—no worries about battery life because the phone is powered from your car's battery.

They have more power, transmitting with 3 watts of power whereas many portables have only 0.6 watts. And car phones are harder to steal because the transceiver is usually installed in a hidden area and the handset can be removed or camouflaged.

Of course, the major disadvantage of the car phone is that the minute you leave your car, you've left your phone behind as well. You can purchase an optional transportable kit, equipped with a battery, antenna and bag which allows you to carry the handset with you. There's also a transmobile kit that allows you to easily move the car phone from one car to another.

Car phone features

Here is a sampling of some of the more common features available:

Automatic volume controls

Hooking the phone up to your radio will automatically turn the radio off when you are receiving or placing a call.

Call summons

You can hook up the phone to your car horn. If the phone rings while you're out of the car, your horn honks. Fortunately for your neighbor's nerves, you can switch this feature on and off at will.

Hands-free answering

Your cellphone will answer the call after an established number of rings and switch to speaker. Another similar feature is "Any Key Answer," which lets you answer the phone by punching any key on the handset. This is certainly safer than having to take your eyes from the road and hunting for the SEND key.

Hands-free dialing

Using voice recognition technology, you program the phone with a list of commonly called numbers and give each number a code name such as "Home" or "Office."

Hands-free operation

This is similar to the speakerphone feature on a land phone. An important safely device, it allows you to talk on the phone while leaving both hands on the steering wheel. It works with a small microphone clipped to your windshield trim and a speaker. A good installer will hide all the wires behind the car's trim and upholstery.

Multiple NAMs

If you regularly drive long distances, get a car phone with multiple NAMs so you can register your car phone with several different cellular providers.

RJ-11 jack

If you travel with a computer, you'll find this feature handy because it lets you plug in a modem or portable fax machine using a modular phone plug.

Theft alarm

Once the phone is powered up, the user must enter a secret code within a specified time. If the code is not entered, the phone silently places a call to a security number and notifies the called party that the phone has been started by someone without the code.

Voice recognition

Some of the more expensive cellular phones can be operated by a set of voice commands. You set your phone into speaking mode which causes the phone to "read back" each number as you press it. This is great for confirming that you're dialing correctly.

Antenna choices

If you opt for a car phone, you'll need an exterior antenna. There are several choices:

* Glass-mounted

* Roof-mounted

* Trunk-mounted

* Removable

A glass-mounted antenna is made up of two parts—one glued outside your rear window, the other glued inside and connected to the cellular phone by cable. This type of antenna is nice because you don't have to drill a hole in your car to install it and it's less obtrusive. On the down side, some users report that the windshield defroster causes interference.

A roof-mounted antenna provides the most powerful reception—mainly because the antenna is higher. But roof-mounts are not very popular—mostly because they need to be lowered before entering most garages. Also, the installation is a bit more expensive because the installer has to drill through the roof and the ceiling lining.

Trunk-mounted antennas are the most popular today. Because the cellular transceiver is usually in the trunk of the car, this antenna keeps signal loss to a minimum.

Removable antennas are designed for particular mounts— your trunk, roof or even the side window of your car. They attach with powerful magnets, a clip or a suction cup. The advantage? You can move the antenna from vehicle to vehicle. It works with convertibles, rental cars, tractors, even a baby stroller. And, if you sometimes have to park in areas that suffer from vandalism, just detach the antenna and take it with you.

In addition to the location of the antenna, you can opt for a cordless or corded antenna. If you are using a transportable phone or car phone, you'll need a corded antenna; if you use a portable phone, you'll need a cordless clip-on or glass-mounted antenna.

You'll also have to select the antenna power. You have basically three choices:

- **Quarter wave antenna** (AKA Unity Gain Antenna). This is the original type of antenna that originally came with every mobile cellular phone.

- **3dB Gain antenna**. This is the most popular choice today. It is designed for metropolitan areas where cell sites are numerous.

- **5dB Gain antenna**. This is best suited to rural areas and open spaces where cell sites are less frequent. This is also an appropriate choice if you're using a mobile phone in a truck cab or other tall vehicle. It reaches greater distances than the 3dB.

Transportable phone

These phones run on portable batteries and weigh between two and six pounds. Transportables are more powerful than portables—up to three watts of power. This helps maintain the signal when reception is weak.

You'll often find transportables phones packaged as "bag phones" with a separate handset, transceiver, battery pack and a sack to lug it around with you. A sleeker alternative builds the phone right into a special briefcase for that James Bond look. Prices start at around $275. You can rent a bag phone for as little as $10 a month.

You can also purchase a transportable kit that allows you to take your car phone with you. Another variation is the trans-mobile or car-to-car kit which consists of a bag, cigarette lighter adapter and often an antenna. This allows you to move the phone from one car to another without having to cope with a lot of wiring.

Laura's luggable

Laura Lee Lemmon carries a cellular phone wherever she goes. She owns and operates Equine Calling Cards, which produces brochures and advertising for the horse industry. Much of her business involves driving over country roads and across long distances to scattered fair-grounds throughout southwestern Ohio. Delays are common and directions can be inaccurate. That's where the cellular phone pays for itself. According to Lemmon, "My phone has helped get directions when I've been lost on the way to a client's farm, and saved my sanity during a very busy season last year."

Laura uses a transportable Okiphone and a service package from Cellular One. She pays only about $30 a month for cellular service. Adds Lemmon, "I especially like the customer-service touch of being able to call and say, 'I'm on my way.'"

Buying a cellular phone

Shopping for a cellphone and choosing a service agreement can be a bewildering experience. Here are some issues to discuss with your salesperson:

❑ **Service area covered**
Ask to see a coverage map to see if the areas covered match your needs.

❑ **Roaming charges**
Ask what roaming charges apply if you use your phone outside your service area.

❑ **Talk/standby time**
How many hours or minutes of calling do you get before the battery gives out? Whatever you're told, divide by half.

❑ **Test drive**
Be sure to make test calls of all the phones you're considering. If possible, test the phone near a tunnel or inside a steel building.

The cellular marketplace is highly dynamic; new features and services are added weekly. A excellent guide through the maze is *Mobile Computing & Communications Cellular Buyer's Guide.* It is packed with product descriptions, reviews, comparison charts and directories.

> *Mobile Computing & Communications*
> 470 Park Avenue South
> New York, NY 10016
> Subscription Department
> PO Box 52406
> Boulder, CO 80323-2406

Choosing a carrier

First, consider how often you may use your cellular phone and where and when you'll use it. Cellular carriers and resellers of airtime offer different rate packages depending on how and when clients are most likely to use their cellular service. Your carrier may offer up to six rate packages, ranging from emergency phone use only to heavy business use.

Your carrier choices are limited to two cellular carriers per area. You can find them in your local Yellow Pages under *Cellular Telephone Service.* Also, your cellular dealer should have information about the rates and coverage of the carriers available in your area. Some dealers work only with one carrier; others contract with both. You are not restricted to the carrier your dealer suggests. Sometimes the dealer's recommendation is based on which carrier gives the heftiest sales commission—nice for the dealer, but no use to you. So ask around, and see what your friends, customers and coworkers are using.

Ask for information about the rate plans they offer and compare the charges carefully. Here are the average rate plans available in the San Francisco bay area:

Sample cellular plans

Type of plan	Monthly charge	Peak airtime	Off-peak airtime
Security plan	$30	85¢	20¢
Occasional use	$25-50	45¢	20¢
Standard use	$50-80	44¢	20¢
High use	$75-115	40¢	20¢
Power user	$135-200	38-40¢	20¢

Many cellular carriers charge on a per-minute basis. That means that at 61-second call is billed as a 2-minute call. If you have a choice, go for the carrier that charges in smaller intervals (such as six seconds).

Check to see if the carrier has send-to-end billing. This means that you will be billed for all calls (even busy or unanswered calls), based on the elapsed time between pressing the SEND button and pressing the END button on your cellphone. Pick the carrier that does not bill send-to-end.

Be sure to read the small print on your contract. You may find that your nice low rate is available only for three months and then climbs rapidly. Sometimes cellular dealers offer really cheap phones but the usage charges are higher. Airtime will add up quickly. There may also be hidden extras—such as installation or an antenna—that are not included in the advertised price. Be sure to check the disconnect clause in the service agreement. Some cellular dealers impose a whopping penalty for disconnecting early.

Carriers sometimes offer perks to differentiate their service. Some offer statewide or area-wide paging services, warranties, or special rates for the disabled or elderly. Roadside assistance services are popular, too. You might sign up for a lost-driver service that will give you directions over the phone, or a road-

side rescue service that gives jump-starts, locksmith service or
emergency gasoline.

Accessories for your cellular phone

Tired of waiting while your cellular phone battery recharges?
You can get a quick charger and battery conditioner that has
your NiCad batteries ready for use within an hour as opposed
to about 10 hours with a conventional charger. ORA Electronics (818-772-2700) makes such an item.

ORA also makes a battery that silently vibrates when your
cellphone gets a call. You just replace your standard battery
with the VibraRing. Great for those times when you need to
be unobtrusive but available.

You can get a device that turns your cigarette lighter into a
four-socket extension outlet. That way, you can power three
more 12-volt accessories such as your fax modem, battery
charger and portable computer (Hello Direct: 800-444-3556).

Cellular fraud

Cellular companies report that they lose close to $1 million a
day to cloners. Cloners pose as legitimate cellular users, but
actually steal the ID and serial numbers programmed into cellular phones. They then install the stolen numbers into
another cellular phone. Then, all calls made on the cloned
phone are billed to the legitimate customer. The customer
often doesn't find out that his number has been cloned until
he receives a whopping big phone bill.

No one is safe from cellular fraud—even the mayor of New
York has had his phone cloned! Cloners lurk at airports, business centers, and other public places. Using radio wave scanners, they lift phone numbers and ID numbers off the air.
They also buy phone IDs from dishonest service personnel
such as car washers, valet parking attendants and the like.

Cloners either peddle cheap long-distance phone calls to one-
time users or sell the phone as a "lifetime" or "magic" phone
and promise the buyer that they will never again have to pay
for long-distance service. Many cloned phones are sold to

criminals—drug dealers and smugglers; others go to unsuspecting folk attracted by the lure of a "no-bill" phone.

How can you tell if your phone has been cloned?

If you experience frequent interrupted or dropped calls, you may be bumping into the person who has your cloned number attempting to make or receive a call at the same time as you. Because both calls use the same cellular channel, the calls "collide" on the air and the later call knocks the earlier one off the air. Don't just assume that these dropped calls are caused by bad equipment or an overloaded system. You should report these instances to your cellular company, which is very interested in stopping fraud.

Carriers are fighting back and installing sophisticated clone detection systems. Some monitor each subscriber's calling patterns and, if they detect a significant change, investigate. Others pay extra attention to international calls, and limit international dialing only to customers who request it. One clever method being used by Bell Atlantic Mobile pinpoints cloned transmissions down to a particular apartment, car or office. Then, they just send in the cops and goodbye to that cloner.

A possible future solution involves sophisticated encryption techniques that would switch algorithms and create one-time secret numbers to foil the thieves. Finally, Texas Instruments Inc. is developing a virtually unbreakable scheme that relies on the cellular-phone owner's voice-print.

Safeguarding your cell phone

❑ Report instances of frequently interrupted or terminated calls, frequent wrong number calls and hang-ups to your cellular company.

❑ Report instances of frequently interrupted or terminated calls, frequent wrong number calls and hang-ups to your cellular company.

❑ Avoid using your phone in or near airports, business centers, and other areas with high cellular use.

❑ Don't publish your cellular number on company phone lists or print it on your business card. If someone needs it, write it in by hand.

❑ Don't allow access to your car phone by people unknown to you. When you use a valet service or get your car washed, take your phone with you or lock it away.

❑ Carefully review your bill every month. Even if you have to pay more for an itemized bill, it may be worth it. A clever phone cloner will be subtle in his use so that the additional charges will be less noticeable.

Cellular and your health

In 1993, cellular phone sales plummeted when rumors circulated that cellular phones were causing brain tumors. Brain cancer allegations were broadcast on radio talk shows, television news and repeated online. The concern was that high-frequency electromagnetic waves surrounding cellular phones were causing injury. Here's what *Time* Magazine ("Dialing 'P' for Panic," October 14, 1993) reported:

> "So far, there have been only a few anecdotal reports of brain cancer among [cellular] users. No one really understands the long-term health consequences of holding a microwave transmitter next to your brain because nobody has thoroughly studied them."

Government studies are now underway to determine if cellphones are dangerous. So far, the evidence is inconclusive. However, if you're worried about the possible effects of using cellular, it might be safest to avoid the portable cellphone, where the antenna is near your head.

Cellular aloft

It's almost impossible to use your cellular phone in the air. Though technically feasible, the Federal Communication

Commission has made it illegal to use a cellular phone while airborne because they fear that cellphones will interfere with vital air-to-ground communications. Although the FCC allows cellphone use while a plane is parked, FAA regulations allow individual airlines to ban cellular use on the ground. Most do. Someday, these regulations should change but, in the meantime, if you need to make a call in the air, you'll have to use the airline's phones.

Cellular at sea

Can you use a cellular phone on a boat or cruise? Depends on where you're going. If it's in the Caribbean, contact Boat-Phone (800-262-8366) who, for $5 a day, will assign you a local 809 phone number. Calls can be made to the U.S. or Canada for approximately $4 per minute (local calls are a little over a $1 a minute).

If you're just cruising inland waterways in your service area, you can use your hand-held cellular phone. However, if you get far from land, your portable won't have the broadcast power to send your signal to the next cell. You'll need a full-powered installed cellular phone if you range far asea. If you do decide to install a cellular phone on a boat, be sure to get a special marine-quality antenna.

For distant cruises, there's mobile satellite phone service. Contact Comsat (800-424-9152) or Stratos Mobile Network (800-432-2376).

Personal satellite communication

Looking for the ultimate gift for your favorite road warrior (yourself, perhaps?) Consider the COMSAT Planet 1, your own personal satellite communications system. Packaged like a laptop computer, and weighting about six pounds, you'll find the technology to connect to an orbiting satellite and place and receive telephone calls from anywhere in the world. The Planet 1 terminal sells for about $3,000. Calls cost about $3 per minute (COMSAT, 800-316-7049).

Cellular fax

Yes, you can send or receive a fax over your cellular phone service. One way to fax remotely is to plug your portable fax machine or modem into the RJ-11 phone jack of your cellular phone and dial the desired fax number. Or get a phone with fax and modem capabilities built-in. Motorola (800-331-6456) makes one—the Micro TAC with the Portable Cellular Connection. Another option is to buy a PC cellular fax modem. This is a credit-card sized device that connects into a cellular or landline phone using a built-in interface and lets you send and receive faxes into your computer from virtually anywhere.

Caution: If you're using cellular fax in a mobile phone, don't drive anywhere. You need a strong, clear signal to maintain a fax transmission. Analog cellular communications are often electronically noisy, with radio interference and weather conditions messing up your fax transmissions. The newer cellular fax modems now offer error-correction and data-compression formats to improve your odds of getting good fax.

Keep costs in mind. Every minute that you are connected over a cellular service costs from 20¢ to 85¢, depending on time of day and where you are located. Thus a three-minute call to send a fax will cost between 60¢ and $2.50. Though not as convenient, a quick stop at a copier or fax service center would certainly save you money.

PCS - Personal communications services

Communications companies have been shelling out billions of dollars to purchase licenses allowing them to sell the next generation of mobile phone service. Called Personal Communications Services (PCS), this new technology operates on a higher and broader portion of the radio frequency spectrum than current cellular networks. This spectrum allows PCS to handle several times the call capacity of cellular. This should mean lower prices for you.

Because PCS is fully digital, the voice quality on PCS phones is clear and free of interference. And PCS can transmit data as well as voice over the same lines without a modem. Handy!

Some PCS phones work with smart cards which contain personalized information about the user, including calling plan, quick dial numbers and billing number. By using their smart cards, several people in your company could share one PCS phone but still get separate bills. Smart cards can also store information like the ID and telephone number of missed calls, text messages and voice mail notifications.

PCS phones

PCS phones look about the same as cellphones on the outside. Because they require less power than cellular, one very useful feature of PCS phones is their longer battery life—as much as five hours talk time (twice most analog or digital cellphones.) The phones also come with many bundled features such as Caller ID, paging, call forwarding, voice mail, and built-in security.

Coming soon is a PCS phone that will work as a cordless phone when near a base station and as a PCS phone when beyond the range of the base. OKI (800-554-3112) has already introduced such a phone in the analog cellular world.

Other interesting possibilities using this technology are wireless voice mail, wireless e-mail and locator services. If ET had such a phone, he would have been home in a flash.

At this writing, PCS charges average 15% to 20% lower than cellular. However, this may change as competition heats up the marketplace. (One company in Las Vegas is charging only 5¢ a minute!) Many PCS service companies offer the first phone minute free for an incoming call. This is useful because it lets you answer every incoming call without worry of charges. If it's a call you don't want to pay for, you have a whole minute to get rid of the caller.

For more information about the fast-changing PCS movement, take a look at the Celltalk Website. It provides vendor addresses, information about PCS phones and service contract information. The URL is *www.celltalk.com*.

Wireless data

If you want to send and receive e-mail or hitch up to the Internet, but don't have a phone connection, there are several technologies that will let you connect:

- **ARDIS**
  Advanced Radio Data Information Services (800-662-5328) is a packet-switched data network service that provides e-mail, wireless fax, and operator-assisted messaging. Pricing is based on the number of kilobytes you send/receive.

- **CDPD**
  Cellular Digital Packet Data offers faster connections (up to 19.2 Kbps). This technology sends packets of data over the analog cellular network. CDPD is available through your local cellular carriers and works with a wireless PC Card, complete with a tiny antenna. You can use this service to connect to office networks and the Internet, as well as e-mail services.

- **Cellular**
  This works with your normal cellular service and a cellphone equipped with a modem interface. Like cellular fax, analog cellular data transmission is error-prone and costly. A 10-minute call to send and receive e-mail will cost between $2 and $8.50. If you establish a data connection, you must stay put until the data has been transmitted. Otherwise, if you cross from one cell to another, your connection will be lost and you'll have to retransmit.

- **RAM Mobile Data**
  Relies on transmitting stations to send and receive data. Like ARDIS, pricing is based on the number of kilobytes you send and receive. This service is limited to text-based data, such as e-mail and fax. For information, contact 800-726-3210.

- **Ricochet**
  The Ricochet Wireless Network from Metricom (408-399-8200) is the most affordable wireless service. You pay a flat rate for unlimited access, and can use the service for Internet browsing, as well as e-mail access and other standard data services. To use it, you'll need to buy a special Ricochet modem.

Resources

Cellular Travel Guide
5th edition, 1996
Communications Publishing Service
Box 500
Mercer Island, WA 98040-0500
800-927-8800

Cellular Telephones and Pagers: An Overview
by Stephen W. Gibson
Butterworth-Heinemann, 1996

Cellular Buyer's Guide
Mobile Computing & Communications Magazine
470 Park Avenue South
New York, NY 10016
Subscription Department
PO Box 52406
Boulder, CO 80323-2406

8...Voice Mail

One of the most significant problems business people face, especially those with small or solo businesses, is fielding phone calls. How many times have you been deeply involved with one customer, either over the phone or in person, and have to leave that customer to answer the other line? And, unfortunately, sometimes end up losing both? What you need is a dedicated, efficient, cheerful, assistant who works around the clock and never takes breaks—coffee or otherwise. Since that option is unlikely, you'll need some kind of automated message taker—either voice mail or an answering machine. We'll tackle voice mail in this chapter and answering devices in the next.

Unless you've been living on a desert island, you've come in contact with voice mail. Though there are many variations, voice mail is basically a computerized messaging system. It routes incoming calls to private voice mailboxes where callers may listen to an announcement, leave a message, get transferred to a live operator, or request that you be paged, depending on the setup. Callers usually interact with voice mail using the touch-tone keypad on their phones. There are two overall voice mail options: (1) having voice mail equipment on your premises or (2) using an off-premise service provided by a phone company or a service bureau.

A survey conducted by the Voice Messaging Educational Committee concluded that the majority of callers prefer leaving a voice mail message to leaving a message with a receptionist or operator, and that nearly 80% of voice mail subscribers feel that voice mail improves their productivity on the job. They found that voice mail systems were rated second only to the telephone as an invaluable communication system, beating out fax, memos, letters and e-mail. The survey also reported that only 22% of callers who reach voice mail hang up, and the remaining 78% leave a message. The Association of Telemessage Services International reports that phone tag can cost a business anywhere from $50 to $150 per employee per month.

A major voice mail benefit is convenience. Since many voice mail services offer the ability to pick up calls after a certain number of rings automatically, you don't have to remember to turn voice mail on. Because voice mail prompts callers to leave detailed messages, telephone tag is reduced. In fact, the editors of Teleconnect magazine estimate that voice mail technology cuts call-backs by 50%.

How organizations use voice mail

Having a voice mail robot can help your business in a variety of ways.

An apartment manager in my area has a separate voice mail number that he publishes in the local newspaper giving basic information about available properties. Prospective tenants can leave messages. He also set up a separate mailbox with the same telephone number to use as an after-hours answering system which tenants use to report repair needs.

If there is an emergency like a major leak, he instructs his callers to press a certain key on their telephone keypad, paging him automatically. Using a call forwarding service supplied by his local telco, he forwards his office phones to voice mail when he is out of the office during the day. That way, he doesn't miss any important calls.

Many school districts have implemented voice mail. Some set up "homework hotlines," where students and parents call to get missing assignments. Others use voice mail as a bulletin board. Callers can get information about lunch menus, after-school activities, sports events.

Public agencies use voice mail as a part-time receptionist to field phone calls and take messages when lines are busy. Transportation agencies have set up street closure and road conditions information boards; animal protection organizations have set up lost pet locators; recreation departments use voice mail as an events hotline and/or event reservation system; and public safety bureaus set up disaster hotlines to be used in case local phone communications are down.

Businesses of all sorts employ voice mail as a useful communications tool for after-hours messaging, 24-hour order entry, market research, salesperson check-in, product information hotlines, prescription renewals, and real estate listings. Many newspapers have added voice classified ads to their services. Restaurants use voice mail as a backup for reservation call overflow. The list goes on.

Voice information line

Steve Burt, a Florida-based businessman, operates Resume House, offering a complete line of job-hunting services from his home office. Steve's voice mail setup contains a series of voice menus where callers can hear a complete description of his services including resume preparation, cover letters and employer mailing lists. His outgoing message prompts callers to leave a message only if they are still interested.

Since adding voice mail, Steve has noticed an increase in business and fewer hang-ups. "Some of my customers actually request an appointment just from hearing my informational message. That never happened before."

Steve reports that voice mail is extra useful when he must spend a lot of time away from his office. "With voice mail, I can retrieve my messages and return calls with little delay. It is almost like having someone in my office to handle my phones for me."

Basic voice mail features

Voice mail systems and equipment do not all come with the same feature set. Here are some of the more common features:

Caller message controls

Lets callers review and re-record messages before they are sent. They can compose, playback, erase, repeat and edit messages.

Date/time stamp

Each message is automatically stamped with the date and time it was sent.

Future delivery

Send yourself a message as a reminder for an important meeting or other event. *Caution:* These messages take up disk space, although they don't show up in your mailbox until the date you specify. Use sparingly if your message capacity is limited.

Group broadcast

Allows you to send the same message to one or several mailboxes at the same time. Useful for notifying a work-group about the time and/or location of a meeting. A real time-saver. Sometimes called group lists or distribution lists.

Guest mailbox

Allows mailbox owner to set up temporary guest mailboxes. Can be used for important visiting clients, temporary workgroups, etc.

Mailbox extensions

Create one or more extension mailboxes so that you can route calls into separate private mailboxes for different departments or individuals who share a single line. Each extension can have its own outgoing message.

Mailbox security

Ability to protect the privacy of mailbox messages with a personal password. Some voice mail systems provide additional security against hackers by locking your mailbox shut after a series of unsuccessful password attempts.

Messaging

Compose and send messages to other persons who are on the same voice mail system. This can save on toll costs and allow you to get the jump on your workday by responding to messages when it's convenient for you. You can record a message when you call in for messages and send it to any other mailbox within the system. Depending on the reach of your system, this could include your entire service area or even the entire state. This feature is often available with off-premises voice mail systems.

Message annotation

Add voice notes to a message and save the message along with your comments as a reminder, or forward the message to one or several people who share the same voice mail system.

Message capacity

Maximum number of messages that can be stored or total message length.

Message copy and forward

Lets you copy a message, add comments to it, and send it on to one person or a group.

Message playback controls

You can pause, skip forward, skip backwards, skip to a new type of message (e.g., new saved, erased), repeat date and time stamp, back up to beginning of a message or the beginning of all messages, jump ahead, etc. These controls allow you to orchestrate the pace of receiving messages; you don't have to furiously copy down messages at the speed of actual speech.

If you are using an off-premises voice mail system, you control message playback by entering one- or two-digit commands from your telephone keypad. On-premises systems often control message playback through commands typed from a computer keyboard.

Message reply
Ability to send a reply back to the message sender immediately.

Message retention period
Length of time that the system retains stored messages before purging them automatically.

Message reviewing options
You may be able to quickly scan all messages and listen to urgent messages first. Messages may be saved, erased, replied to directly back to the sender, or forwarded with or without comments.

Message scanning
Lets you quickly scan through a stack of voice mail. Includes the ability to hear an "envelope" description that lets you know who sent the message and when.

Message sending options
Messages may be marked urgent or private and routed with normal or future delivery.

Message waiting notification
Some off-premises systems use a special "stutter dial tone" that you hear when you pick up your phone. Others light up a lamp on your phone. Some beep you or display an icon on your computer screen.

Multiple greetings
Allows for the storage of more than one outgoing greeting. Can be based on time-of-day, day-of-week, or user can specify when alternate greeting should be played.

Outgoing greeting length
Maximum length of your initial greeting.

Outcall notification
Specify that the system page you. You can usually specify a page for every message received or only for messages

marked urgent. Some systems can automatically dial a number you specify and deliver the voice message.

Training/tutorials

Self-tutorials or help-prompts to assist new users become familiar with mailbox usage skills.

Urgent message handling

Callers can mark a message "urgent," placing it first on your message stack. You can specify that urgent messages prompt outcall notification or paging.

Advanced features

Audiotext

Provides prerecorded information to callers, who control the flow of information with their touchtone keypad. Can be used in conjunction with a voice choice tree (See *Automated Attendant*, below).

Automated attendant

A program that answers your phone, plays a prerecorded greeting that prompts the caller to choose a route for his or her call. May have one or many levels that branch out to provide choices. *Example*: Press TWO for sales inquiries; THREE for technical support, etc. This feature is also called voice menu, voice choice, or menu tree.

Call screening

Requires callers to record their name before being transferred to an extension. The recorded name is then played to the recipient, who then has the option of accepting or rejecting the call. If the call is rejected, it is forwarded to your voice mailbox. *Note*: Callers aren't wild about this feature.

Delivery notification

You can be notified automatically (a message is generated and sent to your mailbox) that lets you know when a recipient picked up your message. Alternatively, you can be notified that a message was not picked up within a specified time limit. This eliminates the "but I never got your message" excuse.

Do not disturb

When enabled, sends a caller directly to your voice mailbox without ringing your phone.

Multilingual greetings and system prompts

Provides user and/or caller with choice of language. Useful for minority population customer base and international communications.

Names directory

An automated attendant feature that provides a directory of all the people and departments in your company. Assists callers to find the correct extension and then rings them through or sends them to voice mail. Often this requires only a partial spelling of the last name using the touchtone keypad.

Night mode

Plays a different set of caller instructions after business hours. May or may not take message.

Number of ports

In this context, ports are basically the same as phone lines. When you call in to pick up a message or when a caller leaves a voice message, you're connected to a port. So, the number of ports equals the maximum number of people who can access your system at the same time. Voice mail systems come in various port sizes, ranging anywhere from two to 9,600. Voice mail boards usually handle only one or two ports.

Outside number messaging

Allows you to send a message to a person who is not a member of your voice mail system. After you record the message, the system places the call and delivers the message. If the line is busy or unanswered, the system tries until the message is delivered or a specified number of retry attempts has been made.

Rotary access

Allows callers with rotary phones to access voice mail and leave messages.

Transfer to attendant

Allows your caller the option to transfer out of voice mail and speak to a "living person" such as a receptionist or assistant.

Caution: If you include this option, make very sure that someone is there at all times to answer the call. You don't want to dump your callers into "voice mail jail," where their only recourse is to hang up or leave a message for you in the wrong mailbox—the mailbox belonging to the receptionist or assistant.

Voice recognition

Provides mailbox control when you speak common word commands such as *play, save, repeat, erase, forward, yes* and *no.*

Voice status announcements

Provides caller with a status message regarding call progress. *Examples:* "I'm transferring you to extension two" and "Extension three is on the phone, would you like to leave a voice message?"

Types of voice mail

You have several voice mail options. You can purchase and install a voice mail board or voice modem in your computer, buy a dedicated voice mail system, or sign up for a monthly service offered by most local telephone companies, paging services, and outside service bureaus.

Voice mail in your computer

If you have a spare slot in your PC or Mac, you can install a voice mail board, connect your phone line to it, and load associated voice messaging software. Voice mail runs on your computer either in the foreground or the background. You can record a greeting or series of branching greetings that provide choices for your callers.

Callers can leave messages in one or several mailboxes. These messages are digitized and stored on your disk. You're notified, usually via your screen, that you have a message. Some voice mail software has the additional facility to page you or dial another number and deliver the message. Obviously, you have to leave the computer on during the period you wish to keep voice mail active.

Voice modems and boards are quite inexpensive and give you complete control over your system. Boards come with single- or multiple-line capabilities. Many have a built-in fax modem, audiotext and auto-attendant capabilities. Mailbox capacity is determined by the amount of disk space you have available.

On the down side, unless you love tinkering with computer hardware, you might not enjoy being responsible for installation, fine-tuning and maintenance chores. Voice processing uses a lot of compute power, so most users suggest that you run voice mail on a separate computer. Otherwise, if you plan to work at the computer and run voice mail at the same time, voice quality may be affected. And, of course, power outages and brownouts will take your system down. You'll need to allocate a lot of disk space to store voice messages. One hundred megabytes for an afternoon's worth of calls is not unusual.

If you have an extra PC (386 or better), and are willing to dedicate an incoming line exclusively, look into a multi-function system like PhoneOffice (Edens Technology Corp., (714-641-1235). Windows-based PhoneOffice combines three separate functions in one system: fax-on-demand, auto attendant and voice mail. Incoming calls are switched to the correct function automatically. You can create up to 999 voice mailboxes and forward messages from one mailbox to another. Pager notification, auto announce, and voice tree capability are included.

All kinds of companies make computer-based voice mail systems for PC DOS, Windows and Macintosh computers. Just visit your local computer store and browse the modem section.

Dedicated voice mail system

Dedicated voice mail systems provide plug-ready voice messaging and other voice technologies that run on stand-alone computers. All you do is plug it in, attach your phone lines, record outgoing messages—and you're in business. These systems are designed to work with PBXs and sometimes, with key phone systems. However, because they are proprietary systems, you may have difficulty integrating dedicated voice mail with the rest of your communications system.

Many dedicated voice mail systems are too costly for the small organization to consider. However, if your voice messaging needs are complex, or you need to support two or more lines, you might look into one of the following: Repartee (Active Voice, 206-441-4700); Infinite Whisper (Vodavi, 800-843-4863); TeleFriend (Future Voice Technology, 510-814-0472); Callstar (Microlog, 301-428-9100); and Resound (Digitcom, 310-584-0750).

Voice mail service

Almost 75% of calls go unanswered, according to a recent AT&T study. There are a variety of reasons: The person called is either on the phone, not there, or unavailable. At that rate, a small operation may be losing a significant amount of potential business. Because a voice mail service provided by a phone company or service bureau can answer any number of calls simultaneously, a higher volume of messages get through, resulting in increased sales and improved customer service. This is probably the most significant reason for choosing voice mail service over other voice mail options.

If you choose voice mail service, your phone can be answered automatically even when all your lines are busy. Overflow calls are forwarded (using standard phone company call forwarding services) to the voice mail computer located at your service provider's premises. You specify when voice mail should answer your calls. It can answer when your are unable to take the call after a specified number of rings and/or if your line or lines are busy. Messages are stored on the voice mail machine. Voice mail subscribers call a special phone number and key in a password to pick up messages from any touch-tone phone.

Voice mail service firms have a wealth of built-in features that can make your voice mail look and sound as professional as those used by major corporations and other large-scale users. They provide optional features such as paging, outcall notification, transfer to attendant, and the ability to send calls from two numbers to the same voice mailbox. Usually you have a limited allowance of the number and length of messages that can be stored. This allowance can be increased for a higher monthly fee.

The Bell operating companies, as well as many smaller companies in your area, offer voice mail. Prices average around $20 a month for a business line and $4 to $8 a month for a residential line. If you have paging service, ask your paging company if it provides voice mail service. Many do. You might find this to be a cost-efficient alternative.

A budgetary note: You may be billed for calls forwarded to the voice mail box and calls made to pick up messages (depending on your particular type of phone service). This can add up, especially if your assigned voice mail machine is not located within your low-cost calling area and you rack up toll charges every time a call is forwarded to voice mail.

There are two main service options:

- **A separate voice mail number**
  Your service provides you with a completely separate phone number for voice messaging. This number never rings at your location. You can give customers the direct voice mail phone number, or forward your calls to voice mail for after-hours (or lunch time) coverage using the call forwarding service provided by the phone company. This service is also called stand-alone voice mail.

  Calls are answered promptly (usually *before* the first ring); callers hear your outgoing message and leave a message. Message waiting indicators (special tones on the line or a light on your phone) are not activated by new messages. You must either dial in to check for messages or specify that you are paged.

- **Voice mail added to your phone line**
  Voice mail capability is added to your phone line. When

your line is busy or you have not answered within a specified number of rings, calls are forwarded automatically to the desired mailbox at the voice mail service where callers automatically hear your outgoing message and leave a message. You are notified that you have a message either via a lamp on your phone set or a special beeping dial tone on your phone.

West-Virginia-based Intra-State Insurance decided on voice mail after its phone system went down during an electrical storm. "We lost power, and our answering machine was out of commission," explains Allan Hawkins, the company's owner. He chose Bell Atlantic's Answer Call service. The company is now available to customers twenty-four hours a day. An added bonus is the ability for sales people to get their messages even when they're out of the office.

At my office

I've had voice mail service for five years now. I pay the residential rate, which in California is less than $7 a month. Because I have a special rate plan that allows unlimited calls in my local calling area, I pay nothing extra for forwarding calls to voice mail. And, as long as I call in for messages from a local phone, I pay nothing extra for my access calls as well. Though the monthly fee is cheap, over the years it does add up. I'm still happy with my choice because, for that price, I get virtually unlimited incoming message service.

I frequently get messages in bursts—three or four may come in at the same time Monday morning, and another peak right before lunch. Unless I had several lines or several ports, my callers would reach a busy signal and they may or may not decide to call back. I also like the messaging function, which lets me send and receive messages to voice mail subscribers throughout California without having to place a phone call.

I have one workaholic editor who used to call me at 5:30 a.m. Since I work out of my home, this is quite an intrusion. Now she drops a message in my mailbox which I can access when I awake—at a more civilized hour.

Problem solving

Problems with voice mail service and Call Waiting

Warning: The tones generated by Call Waiting notification will play havoc with your voice mail setup if voice mail is programmed to pick up your calls when your phone line is busy. The reason: Call Waiting and the call forwarding feature that forwards busy calls are in conflict. Both Call Waiting and busy call forwarding instruct the phone switch about how to handle calls that reach a busy signal, yet only one of the functions can actually take place. Phone systems have decision trees that tell them which function takes precedence if and when they get conflicting instructions. Well, in this case, Call Waiting will override the busy call forwarding feature. This means that none of your overflow calls will reach voice mail when you're on the line. Depending on the switch type in your central office, subsequent callers will either hear a busy signal if the first two callers are connected or be forwarded to voice mail.

If you have call waiting on the same phone line as voice mail service with the busy call forwarding feature, you have a couple of options:

1. **Cancel call waiting before every outgoing call**

    You can temporarily turn off call waiting before making an outbound call so you won't be interrupted during the call. You do this by dialing * 70 (available in most central office switches). Call waiting is canceled for the duration of that one outgoing call. Incoming calls will forward to voice mail. When you hang up, call waiting is automatically reactivated. If you're willing to do this before every call, this will take care of calls coming in while you're making an outgoing call, but it still won't help you when you've answered an incoming call and another call comes in.

2. **Remove call waiting**

    Call your phone company and tell them to remove call waiting from your line. Do you really need it now that you have the ability to get a message from every incoming caller? Removing call waiting will save you some money, too.

3. **Remove busy call forwarding**

   If you can't part with call waiting, you could ask your phone company to remove the busy call forwarding feature from your line since you can't really take advantage of it anyway. Depending on how the pricing is bundled, you might save some money.

Adding message waiting notification

Some people dislike having to pick up the telephone handset and listen for the message-waiting tone to check for messages, a feature of most voice mail services. There are alternatives. You can get a device that clips onto your phone line to notify you that messages are waiting such as the Message Lite (Technology Arts, 800-600-1778) or MessageAlert (Hello Direct, 800-444-3556). Many phone sets come with a built-in message waiting lamp.

Mac owners can use the YoYo Call Manager (Big Island Communications Inc. (408-342-8300). This software-controlled device plugs into your phone line and your computer, and gives you a message-waiting light, Caller ID information, a phone book, call-blocking capability, distinctive ring, and paging capability. When you get an incoming call, YoYo opens your favorite contact manager, (Now Contact, TouchBASE, ACT!, or FileMaker Pro) so you're ready to handle your caller's needs efficiently.

Screening incoming calls

SoloPoint (408-364-8850) makes SmartScreen, a device that hooks up with your phone line and gives you call-screening capability. To set it up, you plug your telephone line into the jack on the back of SmartScreen, and the line to your telephone set. You'll need to sign up for 3-Way Calling service as well as voice mail from your phone company. Then, you can listen to your incoming call, decide whether to pick it up, or let it go on to voice mail. Your caller never knows you're listening in. This clever device comes with a message-waiting indicator and even has the ability to route your voice mail and fax calls.

Which voice mail option is best?

One significant advantage of service vs. hardware is that there are few up-front costs and much less commitment. You pay only an initial setup fee and a monthly usage fee. You can try out voice mail for a minimal cost and see how it works for your business. Then, if you decide to use voice mail for the long haul and the dollars work out, you'll feel more confident about investing in equipment.

Voice mail service is easier to set up than voice mail hardware. Since you already have a phone line, all you need is to add a call-forwarding feature or two to be up and running. And, since voice messaging technology is changing so rapidly, you aren't shelling out for a system that may be obsolete within a year.

Stand-alone voice mail is an attractive option for seasonal businesses and for organizations that want to establish a presence in another market without having to actually set up an office there. This gives you a separate phone number that does not ring anywhere except at the voice mail switch (and, technically, it doesn't *ring* there either).

Probably the most important advantage that voice mail service has over do-it-yourself voice mail is the ability to receive multiple calls simultaneously. By using the phone company's service, your phone will be answered automatically even when all your lines are busy. Overflow calls are forwarded (using standard phone company call forwarding services) to the voice mail machine located at your service provider's premises.

On the other hand, voice boards and voice mail systems provide you with all the control you'll ever want. You're not limited to a set number of messages or a predetermined number of mailboxes. You can customize the system to meet your unique needs. If you want seven mailboxes with outgoing messages of six minutes apiece, so be it. If, next week, you decide to add five more mailboxes, page your technician when urgent calls come in, and reduce outgoing messages to two minutes, all you have to do is program the changes. Though you pay more upfront for a voice mail system of your own, over time it should pay for itself.

With voice mail service, you only have a limited message allowance. Additional mailboxes cost extra (anywhere from around $3 a month for residential service to $20 a month for business service). It's not easy to permanently save a message, because they have built-in delete timers. Users must call their service to make changes such as the number of rings before voice mail answers. Often those changes trigger an administration fee. And, of course, the bill comes month after month.

Voice mail technology is rapidly changing. If you don't own a system, you can take advantage of the upgrades made by your vendor. If you buy, your system may become too limited for your future needs. This major capital outlay may not be cost-justified unless you have a number of phone lines or a communications-intensive application.

If you buy a system, you have to maintain it. This may be easy or difficult, depending on your technical ability and the design of the system. You also have to house it somewhere and keep it clean, dry and cool. And, if you're adding ports, you may need to rewire,

Compatibility with a key system or PBX can be tricky. The phone companies have done a good job of making Centrex compatible with their voice mail. However, if you have a phone system already installed and want to add a voice mail board or standalone service, it's a good idea to contact your phone system vendor and check on compatibility before making a decision.

If you are shopping for a phone system, check out its voice mail functions. Many PBX and key systems come with optional voice mail capabilities.

If you're considering a larger system, you'll need to determine the number of ports you need. The number of ports controls how many people can access mailboxes at the same time. Remember, a typical voice mailbox is used only a fraction of the time. For example, a four-port system can easily be used by 50 to 100 people.

If you aren't sure what your call demands and patterns are, a good way to find out is to sign up for voice mail service with the phone company. Their systems are enormous—with hundreds of ports—so callers rarely run into busy signals and other contention problems. Then note the frequency of calls coming in at the same time.

Depending on the type of business you're in and your application, you'll probably find that you can get by with one or two ports. If you're still unsure, check with your vendor— they have guidelines and tables to help you determine your needs.

Wildfire—ultra voice mail

Wildfire is an automated secretarial service that acts as an efficient personal assistant. Much more than a voice mail system, Wildfire can return calls to people who leave messages, keep track of your whereabouts, even interrupt a call and whisper the name of an incoming caller. You can then decide whether to take the call or send it to voice mail.

When you call into Wildfire, you're greeted with a pleasant voice that cheerily greets you. Anytime during a phone call, you can call Wildfire onto the line and instruct it to place a call for you, create a reminder message, or add the caller's number to your database. If you give your callers a passcode, Wildfire greets them with "Oh hi!" and can play back a message you've left just for them.

Wildfire works with a series of voice prompts, a feature that is especially useful if you are calling from a car phone, because you don't have to take your hands off the wheel to place calls or look up numbers.

A day in the life of ... Rick Smolan

Rick Smolan is a new breed of entrepreneur—the itinerant executive. Smolan created the "Day in the Life" photo-essay book series as well as *Twenty Four Hours in Cyberspace: Painting on the Walls of the Digital Cave.* His projects can involve complex coordination with writers and photographers in as many as 150 locations around the world. To stay on top of it all, Smolan relies on Wildfire, his electronic assistant.

President of Sausalito, California-based Against All Odds Productions, Smolan coordinates his forces from his

> office of the moment—planes, automobiles, hotel rooms, telephone booths, his home, and even at times—his office.
>
> Working in Vietnam recently, Smolan scheduled his electronic assistant to call him each morning. Once connected, he handled all calls at U.S. phone rates, avoiding overseas hotel charges of about $10 a call. By using Wildfire, Smolan also avoids the $15 per-call fee charged for airline phone access. "I once spent four and a half hours on one phone call," Smolan explains. "I made many many calls through Wildfire but never hung up. I finished a half a day's work on my way to New York." Smolan subscribes to Wildfire service from Virtuosity (310-306-2795), a Los Angeles-based communications company. "The phone bills have gone up but so has my efficiency, so that's a pretty good trade," says Smolan.

Wildfire service fees can be somewhat hefty. You can get basic service for about $40 a month but also pay usage charges of around 19¢ a minute. You can also get an 800 number plan. For heavier usage, there are discount plans. An average user will rack up charges of $150 to $200 a month. But, compared to the cost of a dedicated secretary or receptionist, Wildfire is a deal. This service is well suited for the consulting architect whose office location varies daily, the contractor who works out in the field far from the office, or anyone who spends much time on the road.

Large businesses can buy their own Wildfire server, but you'll need about $50,000 to get started. For the rest of us, there are several companies that offer Wildfire service on a subscription basis. To find out what's available in your area, contact Wildfire (800-WILDFIRE).

What's new?
Voice technology continues to improve. New products and services are coming on the market frequently.

Universal messaging
Look for a single interface such as a screen that provides users with a visual list of all their messages, be they voice, e-mail, fax, graphics or an annotated data file. In addition to consoli-

dating the mail, users will be able to access messages through telephone, PC, laptops, display pagers, and other devices not yet dreamed of.

Voice mail over the Internet
If you have a voice-enabled modem, you can add software that allows you to send and retrieve voice-mail messages via the Internet. Check out VocalTec's Internet Phone Telephony Gateway (201-768-9400), Bonzi Software's Voice E-Mail (805-238-5790), and Mediamail's Internet Message Center (770-242-0492).

Coast-to-coast voice messaging
Several telephone companies, including Ameritech, Bell Atlantic, NYNEX, Pacific Bell Information Services, and Canada's Stentor, have joined together in the Messaging Alliance. They plan to offer a new service that will allow their voice mail customers to send voice mail messages to any other voice mailbox in the system. The goal is to create a system as universal as e-mail. This would mean that a sales manager could send a voice message all to his sales agents throughout the country without making any extra phone calls. Ultimately, all 50 million voice mail users could be connected.

Voice annotation
A few e-mail and word processing systems provide the ability to add voice notes to text and data files. This is great for tele-commuters and distributed workgroups. Just think—someone sends you an e-mail message or uploads a spreadsheet file. You review it, attach your spoken comments directly to the file and send it back. Look for Lotus Phone Notes, Centigram's TextMemo, Novell's GroupWise, and Active Voice's Repartee.

Teleputer
Many Windows and Mac computers now come packaged with an intelligent phone, voice answering, fax and modem as well as a variety of telephone management tools. Some come as telephony-enabled workstations; others as add-on devices. Check out the CompuPhone (Integrated Technology Inc., 800-393-8889), and the Interactive Communicator (800-292-2112) for PCs. If you have a Power Macintosh, you already

have telecommunications capability. Just add an Apple Tele-com Adapter Kit (800-776-2333) and you're in business. Microsoft also offers a phone interface.

Tips for user-friendly voice mail

The key to designing a caller-friendly voice mail system that is effective and easy-to-use is to always keep your caller in mind. Remember how you felt when meeting a new voice messaging system for the first time—was it helpful or infuriating? Did you appreciate the convenience of a logical system, or did you feel as though you just entered an electronic twilight zone? Were you confident that the mailbox owner checked in and picked up messages often? Or did it seem that the owner used voice mail as a shield to keep callers at a distance? If you're new to the voice mail game, or just want to brush up on your skills, the tips on the following pages will help you tune up your voice communications.

✔ **Encourage callers to leave details**
A message that is just a name and number is not much good because you have to call back just to find out what the caller needs. Suggest that your callers ask their questions now and, if they have one-way information, to leave it. A friend of mine adds this as a tag-end to his message:

> If you leave me a detailed message, I'll be able to get back to you with the information you need.

✔ **Replace your regular greeting with an alternate greeting after hours and on weekends**
Many voice mail services allow you to record and store an optional greeting. Changing from one to the other is a simple matter of a few keystrokes. Explain your regular business hours, give instructions on how to leave a message, and, if possible, provide emergency contact information.

> This is Power Plumbers. Our normal business hours are 8 a.m. to 5 p.m., Monday through Saturday. After the tone, please leave your name, number and the nature of your call and we'll call you back in the morning.

✔ **Change your greeting frequently**
Your greeting should let callers know whether you are just briefly away from the phone or gone all week. This serves a two-fold purpose: You're giving the caller an idea about how soon you'll return his or her call (two hours, end of day, next week), and your frequent caller will know that you use voice mail actively. Some people even put the day's date in their greeting.

> You've reached Brenda Copeland at Acme Engineering. It's Tuesday, August 14th. I'll be in meetings this morning but will be checking for messages. This afternoon, I'll be in the office.

✔ **Keep greeting short and businesslike**
Although some people may enjoy a good joke or your comments about the latest ball game, the majority of your callers will resent your attempts at joviality.

✔ **Provide instructions for emergencies**
If your business survives on prompt service and your callers need help fast, give them an option. This could be pressing a key combination to label the call urgent, indicating that all or urgent calls ring your pager, or providing an alternate number to call.

> This is the Computer Clinic. I'm out on an emergency right now. Please leave your name and number and a detailed message and I'll get back to you as quickly as I can. If you need help immediately, press the *four* key after you leave your message and I will be paged.

✔ **Speak in a natural tone**
When recording, never read your greeting message. You'll sound nervous or remote. It's a good idea to rehearse your outgoing message until you sound relaxed and upbeat. You might want to send yourself messages periodically to check how you sound.

✔ **Pay close attention to the audio**
Somehow, my dog always knows when I'm getting ready to change my voice mail greeting. I guess he wants to be included in the outgoing message. If you have potentially

distracting background sounds, close the door or change the greeting sometime when it's quiet. This can be tough in a busy garage or machine shop, but, your callers will definitely appreciate it. And, whatever you do, don't use a speakerphone when recording your greeting—unless you want to sound imperious and distant.

✔ **Provide shortcut tips for regular callers**
I really appreciate voice mail greetings that recognize that frequent callers may hear the same informational message over and over and over again. So, tell your callers how to skip your greeting and get right to the message-leaving part.

> If you would like to skip this message and leave a message for me, press the *pound sign* now.

✔ **Check in often**
This simplifies the task of returning calls and lets you quickly respond to important or urgent calls. Be sure to avoid all possible appearance of hiding behind your voice mail.voice mail:

Creating a successful voice mail menu
A poorly designed voice menu can drive your callers away in droves. If you set up a menu tree for callers, you might want to consider these suggestions to avoid "torture by telephone."

✔ **Limit the number of menus and options**
According to industry experts, callers can't remember more than four choices at a time. Try to reduce the number of levels your callers have to wade through before reaching the information or mailbox they need.

> This is the City Planning and Development Division. For planning, press *two*. For zoning information, press *three*. For engineering, press *four*. For all other calls, press *five*.

✔ **Keep greetings and instructions short**
Fifteen seconds (or less) per instructional message is about right.

✔ **Use a consistent form for instructions**
State the desired action first, then the key to press.

> Say: "For sales, press *two*. For service, press *three*," rather than "For sales, press *two*. Press *three* for service." Always say zero instead of *oh*. Otherwise your caller might press the number *six*, the key with the letter *O*.

✔ **Provide the most important or most frequently requested information first**
If the bulk of your callers want to know your hours of operation, tell them that first before going into your menu of options.

✔ **Provide an escape valve**
Don't run your callers through a long list of options or announcements with no chance for escape. Early on, give them instructions like, "If you've already heard this message, you may skip it by pressing the *pound* key," or "If you know the extension of the person you're calling, you may press it at any time." If possible, provide transfer to attendant capability so the caller can speak to a real live person.

✔ **Be kind to rotary phone callers**
Develop a plan that provides instructions for callers using a rotary phone. According to AT&T, 38.5% of American households are still on rotary telephones. The heaviest concentration of rotary users is in older urban households.

Resources

Books, magazines and newsletters

The Voice Mail Reference Manual & Buyer's Guide
by Marc Robins
Robins Press, 1995.
Suite 6J, 2675 Henry Hudson Pkwy., West
Riverdale, NY 10463
800-238-7130
Excellent sourcebook for anyone interested in purchasing a voice mail system.

Teleconnect
12 West 21st Street
New York, NY 10010
800-677-3435
Crammed full of product reviews and stories on voice messaging and general office telephony. Look for the annual voice mail buyer guide.

Computer Telephony
12 West 21st Street
New York, NY 10010
212-691-1191
This magazine is all about the integration of computers and telephones. Can be pretty technical but usually has at least one article about SOHO (small office-home office) communications per issue.

9...Answering Machines

•••

"Dave, this is Hal, your answering machine. Can we talk?" If your answering machine could speak for itself, it could probably tell you a lot—like, "We've had the same old message since 1989," or "I'm on my last legs; don't you think it's finally time to retire me?"

If you purchased your last answering machine several years ago and are getting ready to shop for a new one, you're in for some nice surprises. The new crop of telephone answering devices have been re-engineered to avoid most of the old bugaboos—like hanging up on a long-winded caller in midsentence or continuing to answer your calls even though the message tape is full. They're smaller, using microcassettes and memory chips to reduce their footprint—an attractive benefit for those of us with limited desktop real estate. Some even mount on the wall to get out of your way. Remote operations are easier, and the special beeper you had to lug around is a thing of the past. Many come with full-featured business telephones and useful functions like Caller ID displays and paging capability. Finally, they come in all colors and varieties, from simple stand-alone one-line devices to multi-function office assistants.

Message media - cassette or chip?

Answering machines have traditionally used standard tape cassettes or microcassettes to record messages and to play outgoing greetings. Some require a special tape cassette called an "endless loop" which costs a bit more than standard tape. Most machines are designed for use with standard 30-minute cassette tapes.

The most flexible devices use two separate media—one for your outgoing message; the other for incoming messages. This design allows callers to leave their message immediately after hearing your outgoing greeting without having to wait while the tape fast-forwards to a blank area.

If you've ever left a message with a machine that had an exceptionally long beep, you were dealing with a single tape doing double duty. The beep is merely cover for the tape operations and gets progressively longer the more messages collect.

Nowadays, more and more answering devices are using microchips that record voice messages digitally. Some machines use chips for both incoming and outgoing messages; others use a chip for the outgoing greeting and tape for the incoming messages. Using chips as a recording medium provides better sound quality. And, if your outgoing message is stored on a chip, your callers will hear a crisp, clean greeting every time.

Although chips are relatively maintenance-free (no parts to wear out), they can't store as many messages as tape can. Current chip capacity ranges from 5 minutes to 52 minutes, depending on the manufacturer. But, since a average message takes about 20 seconds, 20 minutes is probably plenty. Some manufacturers also sell an optional chip so that you can double the message capacity.

Types of answering devices

Although answering devices come with a variety of features, there are really only a few basic types: tape-based machines, digital machines, phone plus answering devices, and multifunctional machines. If you're interested in setting up a PC-based voice mail or answering system, see Chapter 8.

Tape-based answering machine

Few entirely tape-based machines are being sold today, however, if your answering needs are basic, you can get a tape-based machine for under $50. Try to steer clear of machines that use just a single tape for both greeting and incoming messages. These machines force the caller to wait before leaving their message while the machine shuttles back to the appropriate section of the tape.

Digital answering machine

In addition to microchip message media, digital offers other advantages such as instant access to all functions. You don't have to wait while cassette tape rewinds or fast forwards. This can save up to 15 seconds a message. Message retrieval is easier and more user-friendly; you can save some messages, erase others, back up and forward to the start of the next message with the touch of a button. Because the system is not mechanical, there are no moving parts to break. The quality of the recorded messages is better.

A digital answering machine will not work on digital phone lines. Although the interface is digital, it still needs normal analog lines to run. If you are lucky enough to have digital phone service (see *ISDN*, Chapter 4), you'll need special equipment.

Phone plus answering device

Many manufacturers package a phone along with the answering machine. Usually, such models cost more than standalone answering devices. Some people like having both devices in one package. I had one once, and when the answering machine broke I was stuck with this bulky featureless phone. That one went the way of the garage sale real fast.

A good use of a phone/answering machine combination is a cordless phone with a built-in answering device. This way, you can screen calls from anywhere without having to be near the base unit—handy for home-based businesses, workshops, outdoor businesses and the like.

Answering device features

Announcement only

Some models have a switch that lets you set the machine to play a greeting or an announcement but not take a message. This can be handy if you're on an extended vacation or don't want to take messages on weekends.

Auto attendant

Allows callers to receive information, leave a message or speak to someone at the touch of a button.

Automatic cutoff

This turns the answering machine off as soon as any handset on the line is picked up.

Tip: If you have a older machine that doesn't cut off, get an interrupt device (costs about $10). You plug it into the wall jack and plug the answering device into it. Then the second you pick up the handset of any phone on the same line, your recorded outgoing greeting is blocked and you can take over the call without shouting over your greeting.

Automatic reset

After playing your messages, the unit automatically resets itself, ready to take the next call.

Call screening

Also called call monitor. By turning up the volume, you can listen to the incoming message. If you want to talk to the caller, you just pick up the handset.

Caller ID

Displays (or announces) the name and/or number of the person who is calling. Useful for call screening. You need to sign up with your telephone company to get Caller ID service.

Change outgoing message by time of day

Program your answering device to play different outgoing messages during selected times of the day (out to lunch, gone for the day, etc.)

Tip: If your current machine only supports a single outgoing message, try recording several outgoing messages, each on a different cassette, and labeling them "gone for day," "in a meeting," "closed until Monday," etc. Then, simply pop in the correct tape and you're outta there.

Confidential message for caller

This lets you leave a specific message for a particular caller. This can be done by using Caller ID, where you program the phone to play a special message for a particular phone number or numbers. Alternatively, some answering machines have a function that lets you give out a special PIN code to certain callers. When they key in their PIN, they hear the special message. You could use such a capability with important customers or vendors.

Dictation

Lets you use the answering machine as a dictation device.

Distinctive ring mode

If you subscribe to a distinctive ring service from your local phone company (see Chapter 3), you can set a switch that will play a different message for up to four different ring modes. Gives the appearance of multiple lines for the cost of one, plus the monthly distinctive ring service fee.

Earphone jack

Lets you listen to your messages privately.

Greeting bypass

Let frequent or impatient callers skip your outgoing message by pressing a touchtone command. You'll have to include instructions on using this command (often an asterisk) in your outgoing greeting. Your frequent callers will really appreciate this feature.

Greeting length

Some devices require that your greeting be a specific length (e.g., 30 seconds). If you don't use up all the time in your outgoing message, your callers are subjected to a few seconds of silence before they can leave you a message (or hang up in disgust). Be sure your device lets you record an outgoing message of any length.

Mailboxes

Lets you create separate greetings and mailboxes for various departments or staff members. Callers select the appropriate mailbox after listening to an announcement that instructs them to "press one for sales, press two for service," etc.

Message forwarding

Program your answering device to call another number and playback your message(s) automatically. Often requires a security code.

Memo

Serves as an electronic notepad. Lets you leave messages for others in your office who use the same answering device—without having to place a call.

Message counter

Usually a digital readout that lets you know how many new messages are waiting. Some devices have two counters—one tells you the number of messages; the other tells you the total number of calls (messages *and* hangups).

Message only

If you have this feature, your machine counts only those messages that have voice on them. Hangups are not saved or counted.

Message playback controls

Fast forward, rewind, repeat, pause, skip, erase, save. The same capabilities should be accessible remotely. If you have a digital answering device, you'll get even more functions such as instant repeat, play new messages only and selective save.

Message waiting indicator

This may be a flashing light or an LCD display that lets you know that a new message awaits you. If you have more than one message mailbox, each mailbox may have its own message lamp. Some machines even use a voice alert: "Dave—you have a message waiting."

Multiple line coverage

Though you pay more, you can now purchase devices that handle two lines. Be sure to check that these devices can actually take a message for two lines at once. Most of them can only ask the caller to "please wait a moment" until they're done recording the first message.

New message playback only

By selecting this function, the machine plays only those messages that haven't been accessed before. Handy if you tend to save messages or share a machine with others.

Paging

Dials your pager to alert you when you receive a message.

Power failure protection

A small back-up battery device can save your time/day setting, remote password setting, message count, greeting(s), and any messages in the event of a power failure. If you need this capability, be sure to replace the batteries periodically. Some models come with a handy battery condition indicator.

Priority codes

An alternative to call screening. You give select callers a special two-digit code which they enter after the answering machine picks up. This causes the machine to alert you that a priority call is coming in.

Programmable hours of operation

Lets you program when the machine should automatically turn itself on. Avoids the problem of remembering to turn the machine on before leaving the office.

Remote controls

Almost all machines allow you to retrieve your messages remotely from a touch-tone phone. Some devices will also let you turn on the machine from afar (usually by letting it ring 10 or 12 times) and update your greeting. If you live or travel in the land of rotary phones, you might also want to have rotary phone remote capability or carry a tone generator with you. Sometimes ads will mention "beeperless remote." What they mean is that you no longer have to carry a tone generator (beeper) to access your answering machine from another phone.

Security

Answering machines are not famous for security. If you come upon one, all you need to do is punch the play button and it tells all. But you can usually get a bit of security remotely; most machines have a choice of at least 99 security codes; some have 999. The more the better. Get a machine that lets you change your password. And then change it frequently.

Warning: Don't take security for granted. Clever phone fraud experts call an unattended answering machine, crack the PIN code and record a new greeting: "Operator, I'll accept the charges." Then they proceed to charge calls to your phone number. You find out about the switch when your humongous phone bill arrives.

Tape full

Plays an alternate greeting to callers when the tape is full advising them that it can't take a message. Hopefully, you pick up your messages often enough that you don't encounter this problem. Some machines simply stop answering the phone when the tape is full—definitely a bad design. If you have one of these, you can't clear the messages remotely because the blankety-blank thing won't answer the phone. Time to replace it.

Time and date stamp

The machine tells you, using digitized voice, the time and date the call came in.

Toll saver

This is a setting that lets the phone ring longer (usually four rings) if there are no messages on the machine and answer more promptly (often one or two rings) if you have messages. The idea is that if you're calling in from a remote location, and you call in for your messages, you can hang up after the second ring. Thus, you won't have to pay for the call just to find out that you don't have a message.

Two-line capability

This convenient feature lets you take messages for two phone lines on the same machine.

Two-way recording

Allows you to record both sides of a telephone conversation. Useful if you need a recorded copy for some reason. Be sure to get your caller's permission, however.

Variable answering setting

Lets you set the machine to answer on the first, second, third, etc., ring. Some machines have fewer settings; some have more.

Variable maximum message length

You might want the ability to specify the maximum length of a caller's message. This can be anywhere from 30 seconds to 5 minutes, or even 30 minutes. This protects your machine from being filled up by one or two talkaholics. On the other hand, you run the risk of hanging up on an important client who just happened to need to leave a long message. Goodbye client.

Voice activation (VOX)

This technology helps the machine determine when to turn off message recording by detecting when a caller has hung up. This saves tape space (or chip space) and avoids those long blank spots at the end of messages. Unfortunately, some machines equate a few seconds pause or a soft-voiced caller with hangup mode. Be sure to test out the VOX in the store.

Voice-assisted operation

A series of voice-prompts confirms that your commands have been implemented: "Dave, I saved that message for you." Especially useful during remote operations.

Volume controls

You'll want to be able to increase or reduce the playback volume.

Multifunction machines

Competition is fierce in the messaging marketplace and has sparked the development of a variety of combo devices that do a little (or a lot) of everything. Examples:

- *Bogen's Friday, the Personal/Office Receptionist*
  Voice mailboxes, built-in fax/modem switch, call announcer, paging, call forwarding, message forwarding, remote programming and music-on-hold (201-934-8500).

- *Panasonic's Fax/Phone/Answering Machine*
  Cordless phone, send and receive fax machine and answering device all rolled into one (800-PIC-8086).

- *IBM's SoHo Assistant*
  Fax machine, answering machine with voice-mailboxes, scanner and 14.4 data/fax modem (800-426-2968).

What's best for you?

One thing we all know for sure—you need some kind of messaging capability. According to a poll conducted by Bruskin/Goldring Research (*USA Today*, 6/24/96), over 64 percent of Americans have a phone answering machine or service. Thankfully, most people have finally gotten over the phobia of leaving a message with a machine. Now what kind of service or device should you get?

Telephone answering devices are very flexible—you are in control of the programming. You can see at a glance that you have messages waiting, and you don't have to place a phone call to pick them up. Answering machines are usually far less expensive than voice mail. They let you monitor incoming calls, which is helpful if you want to get some work done while waiting for that all-important call. If you need to keep audio records, it's easy to archive messages—although you might get sick of a filing cabinet overflowing with tapes. If you have a digital machine, you no longer have to worry about snarled tapes or scratchy outgoing messages.

Voice mail's biggest advantage is that it takes messages even when you're on the phone so your callers don't have to experience busy signals. Message handling on voice mail is more functional—especially the broadcasting and messaging capabilities. A message on an answering device is just that—you can listen to it, save it or erase it, but you can't send it to another user, add comments, or port it to your PC for later reference.

Answering machines are still not the most reliable devices. In a *Consumer Reports* survey conducted in November 1991, 40% of the respondents reported some kind of answering machine malfunction. And, if security is a concern, voice mail is hands down the best choice.

However, if you need to speak to your callers real-time the first time they call, voice mail is no improvement over an answering machine. You'd be better off with a cellular phone or a follow-me-anywhere phone number.

Voice Mail or Answering Machine
What's right for you?

Take this quick survey. If you find yourself marking yes to two or more statements, voice mail or an answering device with multiple mailboxes is probably the better choice for you.

❑   Your callers complain that you are difficult to reach.

❑   You spend lots of time on the phone providing the same information to callers.

❑   You want to add services but don't have the personnel to handle the phones.

❑   Your callers could most likely place an order once they've heard an informational message.

❑   You engage in more than one business.

❑   You need to have your messages kept completely private from others in your company/household, etc.

Sharing a line

Many business phone lines have to do double duty. If yours must handle a variety of tasks, from fax to voice to answering machine, read on.

Daisy-chaining

If you need to plug in other devices, such as fax machine, phone or modem, to the same phone line, you can daisy-chain them. Here's what to do:

On the back of your answering machine you should find two phone jacks, one labeled LINE and one labeled PHONE. Just plug your answering machine into the wall jack using the LINE port. Then plug another phone line cord into the PHONE port on the back of the machine to attach the next device. You can string up to four additional devices on the same line.

Switching

Depending on what you connect, you may need to add a switch that can direct fax calls to the fax machine and voice calls to your answering machine, for example. Some line sharing switches work with Distinctive Ringing, a telephone company-supplied service that lets you have up to four different phone numbers ring on one line (see Chapter 3). You can specify which phone number should ring at each jack. Simpler switches distinguish between modem and fax calls.

The least inexpensive solution is a manual switch that lets your turn a knob and switch between two devices on a single line. Of course, you'd have to know what type of call you were expecting, no easy feat, unless you're psychic.

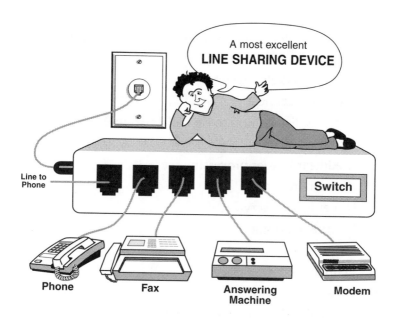

Line to Phone

Switch

Phone   **Fax**   **Answering Machine**   **Modem**

Answering services

Some users still need the human touch they get from an answering service. Answering services are invaluable if you need someone other than the caller to make an on-the-spot decision or immediate follow-up, and you're not available. A well-organized answering service can make your small organization appear larger and more professional.

You have three basic options with a service:

1. You can arrange to rent a phone number from the service.

2. You can hardwire your phone line to the answering service and they will pick up your calls if you don't answer within a certain number of rings.

3. You can forward your calls to them by using a call-forwarding feature from your phone company.

The call forwarding approach is the preferred method for most business situations. It's simple, inexpensive and easy to change.

Answering bureaus provide a range of useful services for small businesses such as:

- Alpha and digital paging
- Audiotext services
- Wireless dispatching to on-call personnel
- Fax broadcasting
- Credit card/order processing
- Brochure requests
- Dealer locator services
- Hotline handling
- Mail service
- Cellular alert
- Wake-up services
- Personalized toll-free 800 numbers
- Voice mail
- Reservations

You can retrieve your messages in a variety of ways. The service can fax messages to you, send them by modem to your PC, or you can call in to pick them up.

Answering services are often willing to handle more complex tasks than simple message-taking, such as taking orders, which can be a fairly complex task for any voice mail system. And many people do not like to give their credit card number to a machine, but are used to giving that information to a live operator. Ask your local service for a list of the services it will provide.

Answering service gives the personal touch
Bob Mastin of Aegis Publishing, Newport, Rhode Island, uses a service to handle overflow calls. He employs a call forwarding feature from his phone company, which sends calls to the service if his lines are tied up or if no

one answers after two rings. Mastin chose a service over voice mail or an answering machine because "neither are very friendly or secure ways for taking an order. My customers aren't comfortable leaving credit card information with a machine. They want to talk to a human being." Mastin pays a flat rate of $75 a month for up to 75 messages. An average bill runs between $80 and $100 monthly, and that's for 24-hour service. "That sure beats having to hire a secretary or an order clerk," Mastin says.

When planning an out-of-town business trip that involved several publicity events, Mastin anticipated a potential spike in order calls, one that his local answering service was not equipped to handle. He therefore arranged to forward his 800 number to an order fulfillment service, specially geared to take telephone orders. You can find such companies in trade publications or directories relating to the Direct Marketing industry.

What's new?

Portable answering machine

The Motorola Advisor Message Receiver may well mark the future direction of message handling. The size of a pager, the Advisor has a small screen which allows you to read short messages. Voice messages reach you two ways: (1) Callers can leave a voice message with an operator who types the message and transmits it to your Advisor. Or (2) you can install Advisor software on your PC and an office assistant can type in messages. These messages are then sent, via the radio airways, to your wireless unit (800-331-6456).

Computer-based answering machine/fax/ phone

Symantec's WinFax PRO provides just about everything you could ask for in an integrated communications device. You install a voice-capable modem in your computer, add software and voila! You have an answering machine, multiple mailboxes, auto attendant, fax-on-demand, fax/voice switching, computer phonebook, Caller ID capability, full-duplex speakerphone, fax send and receive, universal in-box, Internet access software, e-mail and paging. You'll need Windows 95 or Windows NT for this one (800-268-6082).

Tips for leaving an effective message

Since 75% of the nation's largest firms have a voice mail system and almost all the rest have some sort of answering device, the chances that you're going to have to leave a message with a machine have risen dramatically. So, even if you hate talking to a machine, you're going to have to do it. Here are some tips to help you leave the kind of message that will reduce phone tag and increase the odds that your call will be returned.

✔ **Organize your thoughts before calling.**
Be brief and to-the-point. Skip the small talk. If necessary, make brief notes or even rehearse before you call. If you need a response, explain when is a good time to reach you.

✔ **Say your name in the first 10 seconds.**
Most voice mail systems and some answering machines have a scanning feature that allows the recipient to play only the first few seconds of the message. You want to let the recipient know who's calling.

✔ **Start your message with a "headline."**
Explain the subject of your message. This helps the recipient decide whether to listen to the message now or save it for later.

✔ **Give your telephone number at the end.**
Most voice systems allow users to skip to the last 10 seconds of the message because the majority leave their phone number last.

✔ **Always provide your phone number.**
Unless you are absolutely sure that the recipient knows your phone number by heart, provide your phone number. This saves the recipient time because (s)he doesn't have to look up your number and increases your chances

of a callback if the recipient is out of the office and may not have access to phone book or Rolodex.

✔ **Speak slowly when leaving numbers or technical information.**
Doesn't it drive you nuts when someone leaves you a perfectly clear message and then races through the phone number at the very end so you have to play the message over and over to get it right? So, even though you are bored with hearing yourself say your phone number, resist the temptation to hurry through this part.

Resources - Product comparisons
Not sure which answering machine is for you? *Consumer Reports* does a survey every few years. If you have access to CompuServe, go to the Consumer Reports section (Go CSR) and look under Electronics. There you'll find a list of answering machine features, top picks and ratings. If you click on a model, you're presented with a detailed report. Be sure to scan or download the entire report because the best information is at the bottom of the report. It's only there that you learn which model's call counter resets to zero after a power failure and which automatically discards messages after playback unless you consciously save them.

10...Paging
• •

Home-based and small organizations are faced with a catch-22 situation. Their personnel often must be out of the office to perform business tasks, while at the same time they need to be in the office to field calls and maintain customer contact. Too many tasks? Too few hands? One way to handle these conflicting responsibilities is to take advantage of wireless communications.

Industry studies indicate that initial business calls have an average completion rate of less than 20 percent. Having a pager or other wireless communication device increases the probability of call completion to 90 percent. Those figures can translate into more business for you, and they free you from waiting by the phone or playing endless telephone tag.

The are several categories of wireless communications devices, classified by the transmission technology and/or the frequency used. These include: cellular, which connects you to the wired telephone network, using a system of transmitters and receivers; paging, which broadcasts messages over a wide area; wireless data networks, which use a radio modem to transmit information; and PCS (Personal Communication Services), which employ a personal base station to connect to existing landlines.

Cordless phones, though they use wireless technology to handle the connection from the base to the handset, are not technically classified as wireless communications. You'll find information on cordless phones in Chapter 5. Chapter 7 discusses cellular and PCS phones. This chapter will focus on paging, the most useful and affordable wireless technology.

Pagers have come a long way from the intrusive beepers of yesteryear. Today, you can get a pager that vibrates silently, plays a melody, or quietly winks to let you know a page has come in. Pagers come in a variety of styles including credit-card sized pagers, belt clip and ankle clip models (for athletes and joggers). Some pagers hang from a lanyard around your neck; others double as fountain pens or wristwatches. They even come in jazzed-up colors like vibra pink, bimini blue and neon green.

Numeric pagers show the telephone number of the party trying to reach you. Alphanumeric pagers display short messages, often saving you the time and expense of having to make a call to get much-needed information. Two-way pagers give you the option of sending an answer. Voice pagers go one step further and actually speak to you. Today, pagers are often used to notify you of more than just a message. They deliver electronic mail (e-mail), voice mail and even provide information services such as news feeds, weather and hourly stock quotes. Some PDAs (Personal Digital Assistants), palmtops and laptop computers also have built-in paging capabilities. Pagers can even help you if you're lost; there are services available that will send you detailed directions to help you get where you're going.

Paging prices are quite reasonable. You can get numeric-display local or regional paging service for 5¢ to 7¢ a page (message). Alphanumeric paging costs two to three times that.

According to the Personal Communication Industry Association, approximately 48 million Americans carry pagers. A study conducted by BIS Strategic Decisions reports that nearly one out of three cellular subscribers also carry a pager. You see all sorts of people using pagers. Here's a small sampling:

- **Courtroom lawyer:** uses a silent alpha pager in court to communicate with members of her team
- **Real estate agent:** carries an alpha pager to receive listing information on the fly
- **Nurse:** has a silent beeper that alerts him when a patient needs assistance
- **Reporter:** totes a PDA with paging capability to get heads-up information on late-breaking leads
- **Field service agent:** carries a numeric pager to stay in touch with his office
- **Sales manager:** downloads and broadcasts the latest price list to her field sales force

Pagers help you keep in touch with your clientele. Take Pat Rentsch, a California-based telecommunications consultant, for example. Pat's customers are located throughout the United States, and when they have a problem, they need him right away. Since he is often out of the office working at a client's site, Pat has voice mail for routine messages and a pager for emergencies. His outgoing voice mail message instructs callers to mark the message "urgent," and he is paged automatically.

He pays $15 a month for statewide paging service. When his pager alerts him, the client's telephone number is displayed. The service provides him with a Motorola Bravo numeric pager and a monthly allowance of 250 pages, working out at about 6¢ a page. Additional pages cost 15¢ each. According to Rentsch, it has been money well spent.

A painting contractor in my area uses alpha paging to keep in touch. Because most of the job sites he works on have no phones, answering a page can be difficult. If a potential client needs him to drop by for a quote, the client can leave the address with a service that types the message and sends it to him. The system pages him, displays the message, and he can check out the client's job without having to call back to get the details. This way he doesn't miss out on opportunities to bid on jobs, and he avoids the hassle of constantly checking for messages.

Paging can be used in very creative ways. For example, a trio of paging services from Motorola, dubbed "Diner's Delight," provides the restaurant industry with new paging capabilities. One service gives a loaner pager to patrons waiting for a table so they can roam as far away as two miles. This lets them shop a nearby mall, go for a walk, or hang out in a noisy bar. When a table becomes available, the pager alerts them. Pagers are also used to notify restaurant personnel that an order is ready to be served so they don't have to lurk around the kitchen and, instead, can assist customers. A third paging device, hidden in the condiment caddy, lets patrons discreetly call for their waitperson's attention.

SkyTel offers a service called AutoLink that will unlock the door of your rental car should you lose the keys or lock them inside. You call a toll-free number, provide the PIN number to your car, and the system pages your car and instructs it to open up. In California, an enterprising windsurfer offers a paging service that monitors wind speed and direction and beeps you when the wind conditions are right for a wild ride. There are even paging services that send wireless messages to electronic signs, updating them all with a single page. Travelocity offers a paging service that alerts busy travelers to last-minute flight and boarding gate changes.

How it works

You'll need both a paging device and a paging service. When you sign up with a paging service, you specify the geographic area where you wish to receive pages. This can be as small as a portion of a city or as wide as several nations. You also sign up for the type of paging you wish—basic beeper, numeric (which sends you the telephone number of the calling party), alphanumeric (which can send a short message) or two-way. In addition to a service contract, you'll need a pager or PDA device equipped as a paging receiver.

A caller leaves a message for you via a paging operator, voice mail or other system. The paging information is entered into the system's computer, which broadcasts the data (that could be a beep, a telephone number or a short message) over every part of its covered territory using radio towers and/or satellites to deliver the signal. If your pager is on, it "catches" the broadcast and your pager alerts you.

Once you have been beeped, you call in to your paging or voice mail service to pick up your message or, if your pager has display capabilities, read the message displayed on your screen. Only you (or someone with your security code) can pick up your messages. All other pagers within your area are locked out from the data being sent to your pager address.

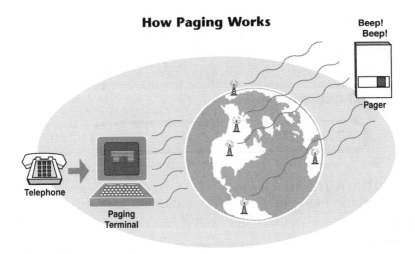

How Paging Works

Beep!
Beep!

Pager

Telephone

Paging
Terminal

Pagers are very reliable communications devices. They were originally designed for doctors and others requiring emergency communications. They use redundant and overlapping broadcast systems and a powerful signal that ensures that the message gets through. Pagers often run for over a month on a single set of batteries and, due to the broadcast nature of the signal, can reach spots other systems can't, such as the interiors of steel office buildings, tunnels and subways. Currently, paging systems can reach over 90 percent of the United States, with more localities being added each month.

Types of pagers
Pagers come in several varieties:

Plain old beeper
This is sort of a radio-controlled doorbell. The pager buzzes or beeps, often quite loudly, to let you (and everyone within 100 feet of you) know that you've got a page. If you dislike the sound, you can get a beeper that vibrates silently. You call your pager service to pick up the message.

Numeric pager
Displays the phone number you should call. If your paging service is connected to voice mail, the number displayed will be your voice mail access number. If your callers directly page you, the number displayed will be the number they keyed in using a Touch-Tone phone.

```
212-555-5873
```

Alphanumeric pager
Displays a brief message (which could also include a phone number) or coded information such as a sender code.

Messages are often limited to from one to four lines. This type of pager reduces the number of return calls you have to make.

```
PLEASE BRING MORRISON FILE
WITH YOU TO THE 2 PM STAFF
MEETING TODAY
```

Two-way pager
Displays a brief message with a choice of responses. These may be preset responses, such as *Yes* and *No* or choices entered by the sender.

> PLANNING MTNG CANCELED
> WHEN TO RESCHEDULE?
> ❶ THIS FRI AT 1OAM?
> ❷ NEXT TUES AT 3PM?

You answer the message by choosing one of the available responses, and the pager transmits a signal back to your service.

Voice pager
Combined with voice mail, collects, stores and plays back voice messages over the pager. Motorola's Tenor pager, for example, can accept and store 16 voice messages. Carry a voice pager as a portable answering machine.

Personal communicator
These are multifunction devices, which may perform calendar, memo pad, calculator and scheduler functions as well as paging and e-mail. These devices go by many names such as PDA (Personal Digital Assistant), PIC (Personal Information Communicator), and PCS (Personal Communications System). Many PDAs—including Casio, Sharp, Sony, and Apple Newton models—support paging, either as a built-in function or as an option supplied by a PC card device. Personal communicators often will allow you to send and receive longer messages, which you scroll to view and which may work with a communications software package located at your home office. See *Paging Communications Software* later in this chapter.

PC message receiver
Designed to insert directly into handheld or notebook computers equipped with a PC slot. Many of these devices can receive longer messages and, if they have compatible software, you can keep the page as a permanent file in your portable computer. Great for sending important information such as price updates, stock availability and market news. Among the companies selling PC pager cards are MobileMedia, Information Radio Technology, and Motorola.

Rent or buy?

You can buy a plain old beeper for less than $10. Numeric pagers range from $50 to $90; alphanumerics range from $120 for a two-line model to about $250 for a four-line version. You can save money if you buy your pager at the same time as you sign up for a service agreement.

You can also rent a pager from your paging company, an option that the majority of pager users employ today. You can usually reduce your monthly charges by $3 to $5 if you provide your own pager (you may be able to negotiate an even larger discount).

Rental agreements usually provide repair and replacement insurance either bundled in the monthly charge or for a low additional fee. If you own the pager, you either have to carry your own insurance or be prepared to foot the cost of repair or replacement. If you plan to keep your pager for over two years, and are reasonably careful, you'll probably save money by owning your pager. The tradeoff is that you may be stuck with obsolete equipment.

Alpha or numeric?

Will an alphanumeric pager save you money? Or just add to your communications costs? To answer that question, determine how much you would save in phone calls each month if your pages contained messages, not just numbers. According to the editors of *Mobile Computing & Communications* magazine, if you estimate a savings of $15 or more, go with the alpha pager.

Pager features

Alarm clock

You can use your pager as a travel alarm. The alarm rings with a different tone from the regular message alert tone.

Automatic message lighting

When a page comes in, the message is automatically illuminated to insure readability, even in the dark.

Display size

If you have alpha or numeric paging service, you'll need a pager with a display to see the message. Some pagers display just one line of the message at a time, others can display two or four lines. Although the size of the pager may be affected by display size, larger displays are handy because you can read a longer message at one glance.

Message lock

The ability to store a special message (or two) in a secure area that is not erased even if the pager memory fills up or the pager is turned off.

Number of message memory slots

Controls how many messages you can store. Sixteen is standard.

On-off programmability

You can preset the pager to turn itself on every morning and shut itself down every evening at a preset time to save batteries. Great feature for those of us who forget to turn the darn things on or off.

Out of range notification

Your pager notifies you by displaying a code or out-of-range message when you are outside call-receiving range.

Priority override

If you select this optional feature, the pager will notify you of an incoming page, even if you've turned off the tone or vibration mode.

Selectable alert modes

Lets you control how you want to be alerted about an incoming page (tone, vibration, light). Some pagers offer a choice of several different melodies.

Time and date stamping

Keeps track of exactly when each page was sent from the paging service.

Total memory

This controls the number of phone numbers and messages you may store. Memory is usually stated in number of characters and indicates the actual storage space available. Assume about 1,000 characters per typewritten page.

Shopping for a paging service

There are hundreds of companies that sell paging services. Many are resellers that offer services provided by one of the major pager manufacturers. Often, the reseller's price is as good or better than that of a well-known service, so it makes sense to comparison shop. Many paging services advertise in the sports section of the newspaper. You can also find pagers listed in the Yellow Pages under *Paging and Signaling Equipment.*

Paging services offer a bewildering number of service plans and options, which often makes price comparisons difficult. Don't automatically pay the listed price. Pager prices and paging service charges are often negotiable. To assist your analysis, I've included many of the most common coverage plans and features. Hope they help.

Coverage

When you choose a paging company, you must choose a paging plan based on the geographic area of coverage:

Local

Local coverage, the least expensive, is adequate for most needs. You often have the option to add one or more neighboring cities to a local plan. Monthly service fees range from $8 to $15 for numeric service.

Statewide

If you cover a specific sales territory, for example, you can purchase a plan that includes your entire state. Costs 25%-35% more than local service.

Regional

Plan includes your state and surrounding states. For example, SkyTel's plans align with Eastern, Central and

Western time zones (the Western zone combines Mountain and Pacific time zones). Costs are approximately 25% more than statewide service.

National

If you travel widely and often, a national plan may be best for you. National paging plans cost two to three times more than regional plans (prices range from $35 to $100 monthly depending on service package).

International

Paging plans are not yet global, but you can get service in many nations including the United States, Canada, Mexico, Argentina, Brazil, Bermuda, Hong Kong, Indonesia, Malaysia, Australia, and Singapore. More countries are being added as the networks build out and alliances are made.

Roaming

A less expensive alternative to a national or international plan is to purchase a plan that lets you *roam*. You notify your paging company whenever you arrive in a new area. The system automatically forwards your messages to the new area. You can, in effect, get national coverage while paying regional prices.

Paging service features

Alpha message length

Depending on the type of service, messages range from one line (15-20 characters) to eight lines. A few plans allow over five pages of text. Longer messages are usually broken into shorter segments for broadcasting. You usually get a message allowance per month and pay a premium for messages exceeding that allowance.

Broadcast paging

Lets you send the same e-mail or alpha page message to a group of people via pager. *Example*: a financial consultant could send stock market advice to a select group of clients.

Follow-me roaming service

You notify your paging service to transfer your pages automatically to a specified city, region, or country outside your normal paging area. Messages will be broadcast simultaneously in your home area as well as your destination area for the period of time you specify. Much less expensive than nationwide or international paging service. Useful if your travel plans include more than one foreign country.

Message receipt notification

Your paging service can notify you when a message you sent was received. You can also specify that you want to be notified if a message was not received within your assigned deadline. This is a very comforting feature.

Multiple language capability

If you deal with a lot of international callers or callers from specific ethnic groups, you might want to specify the language the paging system uses when playing system prompts. Spanish, English, Japanese and Cantonese are among the languages supported.

Number of paging attempts

Number of times the paging system will attempt to reach you for each message. The standard is two, which is usually sufficient.

Page recall

Lets you dial into a pager system and retrieve stored messages. These could be messages you have never received or messages that you stored for later retrieval. Convenient if you were outside paging coverage area or your pager was turned off for a time.

Personalized greeting

Some paging systems provide the option of creating a digitized greeting in your own voice. When callers dial your pager number, your greeting is played. You can use this feature to update others on your travel plans, business developments, or instructions on how to access you.

Personal 800 number

Your paging company can supply you with your own toll-free number that your callers dial to leave a message and you use to pick up messages. You usually pay a small monthly fee ($2 to $5) for this service.

Personal identification number (PIN)

Your callers will need to know your PIN code in order to have you paged, unless you have a personal 800 number that is forwarded to your pager.

Personal security code (password)

You must enter your password in order to pick up your messages. Try to get a paging service that allows *long* passwords. Many paging companies allow only four digits and this, in my opinion, is too few for adequate security.

Priority paging

An optional service that will retry a single message more times than normal paging will attempt. The standard is two attempts. Useful if you need to be sure that urgent messages get through to you.

Sequential page numbering

You'll see a message number on your pager display. Pages are numbered in the order sent. If the unit was turned off or out of range, this handy option helps you determine if you missed some messages.

Simulcast service

Allows you to specify that all page attempts be broadcast both in the United States and the country listed in your international paging plan. This service is most useful if you frequently travel between two countries and don't want to be bothered setting up special paging arrangements for each trip.

Time-of-day paging

Lets your callers send time-sensitive messages for future delivery, making it easy to communicate across time zones.

Usage security code

You can control who can leave messages for you by requiring callers to enter a special security code. No code, no message. This feature helps you control costs and avoid unnecessary pages. Especially useful if you are paying a premium for alpha paging service.

Voice mail option

You can often get voice mail service, at a reasonable cost, through your paging service.

Voice page copy and forward

You can electronically copy a voice mail page, add comments, and forward the message to another person. *Example:* A real estate agent receives a voice message containing a property offer, copies the message, adds voice notes about a counter-offer, and forwards the message to the seller's agent.

Service costs

You'll pay a one-time activation fee, a monthly service charge and over-call charges if you exceed your monthly page allotment. Costs vary greatly, based on your usage and coverage plan and the services you select. A few paging services offer fixed-cost plans, which may be attractive if you receive lots of pages. The chart below compares nationwide coverage, including pager rental, from three major paging services.

Numeric Paging - Nationwide Coverage

	Company A	Company B	Company C
Monthly charge	$36	$25.95	$39
Page allotment	200	200	200
Overcharge/page	25¢	25¢	25¢
Average cost/call	18¢	13¢	20¢
Personal greeting	included	$2	included

Security concerns

Now that many pagers have messaging capabilities, they are beginning to be targeted by hackers who change security codes and take over mailboxes. If this happens to you, your security code won't work and you'll be unable to pick up messages. Contact your paging company and it will change the code again for you. To reduce the odds of being a victim, change your password frequently and don't use easy-to-crack passwords such as your name or birthdate.

Paging companies are fighting back against fraud and airwave theft. If you send critical information over a paging network, ask your paging service how it can help protect your data. If the information is highly sensitive, use more secure communications.

Paging communications software

If you have someone back at the office to take a message for you—and you have a personal computer, pager and modem— you can turn the combination into a personal paging system. This is how it works: Your secretary, family member or coworker takes a message for you, types it into a personal computer and sends it, via modem, to a service that alerts your pager. Your pager displays the message sent. Several companies make inexpensive paging software, including Apple, AT&T, ExMachina, Hewlett Packard, and Motorola.

Many regional and nationwide paging services offer e-mail paging, which lets you receive either the entire message or just a heading (usually sender's name and topic). You can also have mail forwarded to your pager from public e-mail systems like AT&T Mail, the Internet, and MCI Mail, and from private systems like Lotus' cc:Mail and Novell's GroupWise.

Two-way radio mail

One of the hottest new communication services is RadioMail, which broadcasts text messages over the airwaves that are picked up by using a radio modem, such as Ericsson GE's Mobidem and Motorola's Info-TAC. What sets this technology apart from paging is the ability to have two-way, real-time wireless messaging. RadioMail allows you to send and receive unlimited e-mail messages and faxes using a portable computer or PDA and a radio modem. Service plans start at about $40 for nationwide coverage.

RadioMail uses a store-and-forward method of messaging. This allows you, for example, to send a bunch of e-mail messages from a ferryboat in Puget Sound and pick up the replies when you next log on to the system after arriving in Seattle. No phone lines are involved. If your modem is plugged in to your computer and your laptop is on, you're connected to the system.

When someone sends you a message, your computer beeps. RadioMail provides all the interfaces and protocol conversions to access the Internet, MCI Mail, AT&T Mail and many other services. Messages can be sent to anyone who subscribes to the RadioMail service and to almost any electronic mail address (800-597-MAIL).

Replace cellphone with radio paging

Vicki Whiteford, who owns All-Ways Courier based in Los Angeles, switched from cellular service to Radio-Mail. Each driver carries a HP 95LX Palmtop PC and an Ericsson GE Mobidem radio modem. According to Whiteford, "We can get messages to and from our drivers whether they're stuck in traffic or in the bowels of a concrete parking structure. If we used a pager, a driver on the freeway would have to find a way to get to an exit and then locate a phone to contact the office."

RadioMail can provide the driver with a complete message with all the information the driver needs to know. If the driver is out of the coverage area, the message waits at a RadioMail gateway until it can be delivered. It won't be lost.

Whiteford points out several reasons why she switched from cellular phones to RadioMail. One was coverage: Radio-Mail messages kept drivers in touch, even in the Mojave Desert, a notoriously difficult area for pagers or cellphones to reach. Cost was another major motivator. "Cellular phones can be shockingly expensive and, if you can't get through, you've got to keep trying," she adds. "With RadioMail, you compose a message, send it and forget it. You know it will be delivered—you'll be notified if there's a problem."

All-Ways Courier uses RadioMail for many everyday business transactions as well. "The result," Whiteford reports, "is hundreds of dollars saved monthly in both cellular and standard phone charges."

PDA pagers

A host of PDAs come with paging capabilities. Among the best are:

*   *Apple's Newton MessagePad*
    Comes with an external fax/modem. You can also insert a PC pager card (800-538-9696).

*   *Hewlett-Packard's Omni-Go*
    Works with a standard PC card or with SkyTel's two-way paging service (800-443-1254).

*   *Casio's Cassiopeia*
    Comes with a PC slot that will accept the SKyTel Messenger and Page-Card for one-way or two-way paging (800-962-2746).

Do you need a pager?

Take a moment to complete this survey. If you answer *yes* to three or more statements, paging would benefit your business.

❏   Are you losing business to your competitors because you aren't able to speedily return calls?

❏   Do you want to able to respond to callers more quickly, but need an economical method?

❏   Are you out of the office a lot, but not near a phone?

❏   Do you need to keep abreast of late-breaking developments?

❏   Do you need to be in touch, but don't require two-way communication?

❏   Do you run a service business such as computer repair, automobile towing, plumbing, delivery, etc.?

❑ Do you worry about not being reachable in an emergency?

❑ Do you waste time (and phone change) constantly checking in?

❑ Do customers, coworkers and clients need to contact you at a moment's notice, regardless of where you are?

What's new?

Internet paging

If you have an alphanumeric pager, check with your paging service to find out how to use Internet messaging service. For example, if you have PageNet service, you provide your customers with your PIN number, and they create an e-mail message on their computer. Then they address the e-mail to *your PIN number @pagenet.net* and the message is sent to your pager. Neat!

One warning: Don't use Internet paging for urgent messages. You can't always count on speedy delivery, especially during peak usage hours.

Beeper watch

You like the idea of a pager, but don't want to carry one around? Both Seiko and Swatch sell wristwatch pagers. Seiko customers can also subscribe to information services such as weather forecasts, financial information, air pollution information, and sports scores. The watches come in many styles and are surprisingly affordable. (Seiko: 800-456-5600; Swatch Telecom: 800-8-SWATCH).

Piggyback pager

Motorola offers a numeric pager built into the battery of your cellular phone. Dubbed RSVP, the pager can beep you even when you're on your cellphone or the phone is turned off (800-775-2913).

Personalized pager disk

ExMachina has an interesting product that bundles nation-wide paging service with paging software. You give preferred callers and customers a ReachMe floppy disk which contains your contact information and pager number. They just insert the disk into their computer, click on an icon and you get paged. An interesting twist on the standard business card (800-238-4738).

On the horizon

The FCC recently auctioned a block of new radio frequencies set aside for new communications services. Expect a wide variety of new paging and wireless applications shortly, as developments roll out. The new wireless networks will heat up competition and should force prices down. According to industry experts, wireless technology should cost about half as much as cellular. Good news for you and me.

Pagers will probably be produced as a single computer chip that can be integrated with a variety of computing and communication devices so that you don't have to carry a separate gadget for each function. Interactive wireless voice mail capabilities over paging networks are in the works.

The line between paging and wireless e-mail is not very clear and is getting less so every month. See Chapter 11 for more information about e-mail.

Resources

Nationwide paging companies

SkyTel
800-456-3333

MobileComm
800-685-5555

PageMart
800-324-PAGE

PageNet
800-PAGENET

Sprint
888-617-6792

AT&T Wireless Services
800-462-4463

MetroCall
800-800-BEEP

ProComm Nationwide
800-2-TURNON

11...Going Online

If you've never checked out an online service, wandered the Web, or visited a BBS (Bulletin Board Service), you're in for a treat. The online world offers something for everyone. You can buy office supplies, order a pizza, advertise your business services, buy or sell stocks, conduct market research, make an airline reservation, check out your competition, or visit with a long-lost friend.

You can check out the latest sports scores, view the winning lottery numbers, get late-breaking news updates and save hundreds of dollars by downloading free or inexpensive software. You can ask for help from experts in every field, find a job, even earn a college diploma. And it's getting easier and less expensive to go online every day.

If you think that the time has come for you to join the online world, read on. This chapter will briefly review some of the national online services and explain how to sign up for service and offer suggestions on how to use electronic mail (e-mail) effectively. I'll provide an overview of the Internet, and suggest some ways you could use the World Wide Web to promote your business. I'll also discuss how to recognize (or better yet, prevent) computer viruses, and briefly discuss file transfer, including how to upload and download files.

An introduction to online services

You'll find a multitude of services online, but they break down into only a few major areas:

News and information

There are all sorts of news services online. A sampling:

- Newspapers like the *New York Times, San Jose Mercury* and the *Chicago Tribune*
- Magazines like *PC Magazine, Time, Home Office Computing* and *National Geographic*
- AP, UPI and Reuters news wires
- Weather reports worldwide, including special reports for aviation
- Full-text copies of U.S. government documents and proceedings

Finance

- Investment news
- Standard & Poors' profiles
- Stock quotes
- Portfolio management & brokerage services
- Interest rates
- Online mortgage calculator

Reference and learning

- Encyclopedias
- *Consumer Reports'* ratings and articles
- Online dictionaries and thesauri
- Demographic information (sales potential, target markets, residential neighborhood market analysis)
- Abstracts and full-text articles from hundreds of newspapers, magazines and newsletters
- Databases containing specialized information such as a private-eye information service, a veterinarian's medical database, criss-cross address directories

E-mail and other communications
- Electronic mail links to almost anywhere (more about e-mail later in this chapter)

- Fax, telex and U.S. mail capabilities

- V-mail (voice attachments to electronic mail)

Shopping and travel
- Eaasy Sabre and the Official Airline Guide—look up flight schedules and make your own airline reservations online

- Car rental and hotel reservations

- Electronic malls containing a wide range of products such as software, office supplies, gifts, flowers, computer supplies, books

- Automobile pricing service, home sales information and more

Forums
One of the more useful features of online services is the opportunity to meet other business people and share ideas—sort of a worldwide chamber of commerce. Forums and special interest groups meet a variety of needs such as:

- Law
- Journalism
- Desktop publishing
- Education

- Real estate
- Home-based work
- Medicine
- Engineering

Each forum has a message board, basically an electronic bulletin board with areas designated for certain types of messages. For example, if you want help selecting a new office chair, you might post a notice in the Working From Home Forum on CompuServe. To do this, you type a forum question offline, log onto the service, select the appropriate forum and subtopic, and drop off the message.

Anyone who visits the forum may read the message and comment. I may get one response, but more likely I'll get 10 to 20 from helpful people around the globe (though most are from the U.S. and Canada).

Subj: BEST OFFICE CHAIR?   Section: Office Hwdr & Softw

FR: J Langhoff 71022,2131

TO: ALL

I'm looking for a new ergonomic office chair and am hoping that I can get it
for less than $300. I need one that has pneumatic lift, no arms, and hopefully,
adjustable back and seat tilt. Any suggestions?

The message will stay on the bulletin board until it "scrolls
off." Because the boards have limited capacity, each time a
new message is added, an old message is dumped. In very
active forums, messages are cycled off in three to four days. If
the sysop (system operator) decides that my message and the
answers to it are worth saving, s(he) will group all those messages into a "thread" and store them in the library where you
may read or download them later.

Computing & software

The online world is full of help forums staffed by experts from
equipment manufacturers and service providers that can
assist you in solving most any computer-related problem. For
example, while working on this book, I discovered something
desperately wrong with a file that contained an entire chapter.
Every time I opened the file, my computer froze. I called
Apple. They said it was Microsoft's fault. I called Microsoft.
They blamed Apple.

This was costing me time and money (Microsoft's help line is
not toll-free and holding times were long) and making me
crazy. Then I remembered that both Apple and Microsoft had
forums on CompuServe. I quickly composed a forum message
explaining my plight and posted it on Microsoft's Word
forum. The next morning when I logged on, I received not
one but four helpful responses. And the first suggestion fixed
the problem.

You can also find tons of shareware and freeware programs
available for a free trial. Shareware works like this: The shareware programmer makes the software available online for you
to try. You can download the program and try it out. If you
like it and plan to use it, you send the software programmer a

small fee ($5 to $25) to register yourself as a user and qualify for updates, user manuals, and other support.

Fun stuff

Of course, there's all sorts of entertainment online including humor forums, cartoons, hobby areas, movie reviews, and games such as chess, trivia, adventure games and the like. I try to avoid these areas because they can suck up online time, but if you're judicious about how long you allow yourself, you may find such diversions a welcome break in your day.

How to sign up for an online service

Online services operate on a subscription basis, and you are billed monthly. Most services have a basic monthly service fee that will provide a minimum number of hours of online connect time and a set number of services. Some also offer a flat fee for frequent users. Additional services (such as research services) often carry a surcharge, ranging from 25¢ to over $1 a minute.

Note: Be sure to find a service where your calls to it are billed as local calls, thus avoiding toll charges. If you live in or near a major city, this will not present a problem. However, some outlying areas do not have local access numbers, and you would incur toll charges when dialing in to the service. Since online sessions tend to be long (I rarely get in and out in less than 15 minutes), phone costs could run up quickly. To keep costs under control, get the fastest modem you can (see Chapter 12 for information about modems).

Most online services offer free start-up kits, containing software and user guides to get you going. They also often provide several hours of free service the first month. This way, you can "test-drive" the service at no cost to you. To get started, just call one of the numbers listed in the Online Sampler, below.

You'll receive a user ID and will be required to select a secret password and, in some cases, an online identifier. I recommend that you use your own name or business name. Your ID will appear on all your mail and forum communications and will form a part of your e-mail address (more about this later in the chapter). Be careful when choosing a name, as some services have no facility for changing your mind later. Though some people use handles like "Wild Bill" or "Miss Kitty," these seem rather silly in a business setting.

An online sampler

America Online (AOL)
America Online is very easy to use and contains a wide range of services for businesses such as the Small Business Advisor, Business Strategies Forum, Hoover's Business Resources, Real Estate Center, the SBA Online and the PC Financial Network (800-827-6364).

CompuServe
CompuServe has been online the longest. It started in the '70s and offers the best research services and professional forums around. There are special forums for public relations professionals, pilots, lawyers, engineers, entrepreneurs, photographers, musicians, and physicians. A favorite with home-based business people is the Working from Home forum (I'm a section leader there—come visit!) (800-848-8199).

With over 1,700 databases, including Knowledge Index, IQuest and the Legal Research Center, CompuServe is the choice of most researchers. One of my favorites is the Executive News Service, a clipping service that will scan the electronic newsways for you, clip articles that pertain to the subjects you've specified, and keep them in tidy folders until you have a chance to read them.

Prodigy
Prodigy positions itself as the online service for families. It has special kid's forums, online chats, news services that include full-color photos, and lots of opportunities to shop and visit online. Prodigy provides excellent financial services and is home to the PC Financial Network, an online brokerage service that features free market reports, and BillPay USA, an online bill-paying service (800-PRODIGY).

Other online services

In addition to the major online services profiled above, there are many other services available. Some of the most popular are:

- **AT&T Business Network**
  An online network specifically designed for business needs (800-222-0400)

- **Checkfree**
  Electronic bill-paying service (800-882-5280)

- **Delphi**
  General-purpose service with excellent databases such as DIALOG (contains over 330 million articles) and services such as the Delphi Librarian that will hunt down information for you (800-695-4005)

- **Fidelity Online Xpress**
  Online trading and investment-tracking services (800-544-0246)

- **GEnie**
  General-purpose service with specialties such as GEnie QuikNews (a clipping service), an interactive Business Resource Directory, and evening round-table discussions on a wide range of topics (800-638-9636)

- **Microsoft Network**
  Currently, you must have Windows 95 installed to access this service (800-386-5510)

- **Lexis and Nexis**
  Full-text information about legal (Lexis) and business (Nexis) news (800-554-8986)

- **Streetsmart**
  Windows-based investment service from Charles Schwab (800-435-4000)

- **Women's Wire**
  Excellent spot for networking with professional women (men are also welcome) (800-210-8998)

Advertising online

Most online services and BBSs (bulletin board services) have rules about where you may advertise. CompuServe has a Classified Ads section, where for a small fee, you may post an advertisement. America Online offers free want-ads to members.

Although advertising is not allowed on individual forums, they offer excellent opportunities for networking. If you see a message in a forum requesting information that you think your business can provide, it's perfectly okay to send a private e-mail message suggesting your product or service. Just don't send a blatant ad.

E-mail

E-mail is simply an electronic method for moving information—whether text, graphics, video or sound—from one person to another over a computerized connection. According to the Electronic Messaging Association, there are somewhere between 30 and 50 million e-mail users today. It's expected that e-mail use will be universal by the end of the decade. Conducting business electronically will soon be the norm.

E-mail is more than just a substitute for a regular postal service (dubbed *snail mail* by online humorists) or fax transmission because, once a message is computerized, you can copy it, store it or forward it to someone else for handling. You can even get an electronic return receipt to let you know when someone opens your mail. Another advantage of e-mail is that it eases communications across time zones. For example, if you regularly trade with partners in the Far East, it's a lot easier to send an e-message rather than get up in the middle of the night to place a phone call.

Electronic mail is especially useful for small businesses because it levels the playing field and lets you compete with the big guys. When you're doing business electronically, there's really no way to tell whether you are a tiny start-up or large firm. Your e-mail address lets you be as accessible as anyone else.

How e-mail works

E-mail is great for collaborative projects. An example: Charlie Swanson runs Edgewater Productions, a San Francisco area film, video and multimedia production company. Charlie stays in touch with a network of writers, designers, and production assistants via e-mail. Charlie and I have worked together on videos and we didn't even meet face-to-face until we were well into our second project. We communicated strictly via phone, fax and e-mail.

Here's how the e-mail messaging worked: First, I wrote a rough draft of the script using my Macintosh word processor. Then I started my America Online software, typed a brief message, and picked Charlie's e-mail address from my computerized address book. Next, I clicked on a file icon to attach a file to my message, picked the file from my directory that contained the script and clicked SEND. The software started up my modem, dialed the online access number, sent the message and logged me off—in less than 30 seconds. That simple.

Later, when Charlie signed on to America Online, his mailbox showed messages waiting. He just clicked on his mailbox, and my message with the attached file popped up. He read the message, saved the file on his portable PC-compatible machine, and began to edit it. When he finished reviewing the script, he typed in his changes and suggestions and sent it back to me via another e-mail message. Using e-mail technology, script updates, rewrites, scene changes, and last-minute casting substitutions were handled almost instantaneously, cutting days on pre-production time and improving the bottom line. Before I used e-mail, I had to messenger scripts to my clients—at a minimum of $40 a trip (unless I could convince them to wait for next day service, something most harried producers were unwilling to do).

E-mail messages are pure ASCII text and contain no formatting. ASCII (American Standard Code for Information Exchange) is the code that allows virtually all computers to talk to each other, regardless of their individual platforms. Because the messages are in ASCII format, they can be sent between computers using different operating systems. This lets you communicate easily with people using Macs, Windows machines, Suns, Amigas, and mainframes. E-mail works with them all. Incidentally, the files attached to e-mail aren't limited to straight text; they can contain graphics, spreadsheets, artwork, software, page layouts, and even music. Just be sure that the person on the receiving end has the appropriate software to open the file or has a file viewer program like Lotus Notes or Adobe Acrobat.

Many solo and small businesses find e-mail more economical to use than regular mail. Even using a extremely slow modem, such as a 2400 bps device, you can send a 200K file (about 50 typed pages) in about 16 minutes. At a faster speed, such as 14.4 Kbps, that same file would take less than three minutes. At standard online rates (about $2/hour), the slow file would cost around 80¢ in online time; at the faster speed you'd be out around 13¢. Best yet, the file would arrive at the recipient's desk within minutes rather than days. If I had printed out the file, and sent it Priority Mail, it would have taken at least two days and cost $3.

There are some e-mail negatives. You must log on and check to see if you have any mail. This takes time and can run up phone charges, especially if you have to pay toll charges. E-mail can pile up alarmingly fast so, if you find yourself with more than you can handle, consider getting an e-mail management program, such as Eudora Pro (Qualcomm, 800-2EUDORA) or E-Mail Connection (ConnectSoft, 800-889-3499) to help you to screen, save, delete or archive messages. Workgroup software such as Microsoft's Windows for Workgroups and Lotus' Notes also come with built-in mail management capabilities.

E-mail can also get stuck. So, if you're used to getting frequent messages, and suddenly the steady stream dries up, don't assume that you've lost popularity. It's possible that the mail server has a problem. How to check? Ask someone to send you a message, or, if you have more than one service, send yourself a message. Call the administrator or help line. I recently experienced this problem with America Online. When the mailbox was fixed, thirty-four messages were waiting for me (some five days old).

How much does e-mail cost?
Most services offer unlimited e-mail. Your charges are based only on connect time. If you compose e-mail offline and just sign on to upload or download messages, you'll handle your e-mail duties in a minute or two. I send somewhere around

ten messages a day and receive about twice that many. That works out to a little over 1¢ a message. What a deal!

Depending on your area, you may also be able to sign up for a free e-mail services such as Hotmail (510-440-8912) or Juno (800-654-5866). You'll have to put up with some advertising. Also, these services currently can't handle attached files.

How to get an e-mail address
Once you have an online service or an Internet account, you are automatically given a globally-accessible address and e-mailbox, which lets you send and receive mail to anyone else who has an e-mail address.

What's your address?
To send a message to another person, you need to know his or her address. You can look up the address online (if the recipient is listed) or call him/her and ask for the address. If you both have the same e-mail service, you just enter the recipient's e-mail address and send off the message.

But, if you have one service (say, Prodigy) and the recipient has another (MCIMail, perhaps), you need to use the person's Internet address. This is just the person's regular e-mail address plus some other codes to direct your message through the maze of networked computers. The Internet address is made up of the following:

Internet suffixes
.com	=	Business or commercial service
.edu	=	School or university
.gov	=	Government agency
.mil	=	Military installation
.org	=	Noncommercial site such as a BBS
.net	=	Network

Seven new extensions are in the planning stages and should be ready by the time this book is in print. They are:

.firm	=	Business
.store	=	Companies selling products
.web	=	Sites emphasizing the World Wide Web
.arts	=	Cultural sites
.info	=	Information services
.nom	=	Individuals
.rec	=	Recreational activities

If you wanted to send an electronic message to the president, you would address it:

president@whitehouse.gov

Note: If you're sending to someone on CompuServe, you'll need to substitute a period for any commas in their address. For example, my CompuServe ID is 71022,2131 and if you wanted to send me e-mail via the Internet, you'd address me thusly:

71022.2131@compuserve.com

Some e-mail services require that you preface the e-mail address with "Internet:" If your e-mail doesn't go through, try adding the preface and sending again. This is what the address would look like:

Internet:71022.2131@compuserve.com

How to find someone's e-mail address
Many Internet search engines, such as Infoseek and Yahoo, offer e-mail address searching. There are also special services, such as WhoWhere? *(www.whowhere.com)* and Four11 *(www.four11.com)* that contain millions of listings.

Is e-mail private?
Not entirely. Anyone who receives e-mail from you can copy it, store it, forward it to others, even edit it. As Oliver North and others learned during the Iran Contra scandals, e-mail may be stored for years as part of a routine disk backup. Therefore, it makes sense to use e-mail prudently. If the information you send must remain private, use an encryption program such as PGP (Pretty Good Privacy; 415-524-6221). If you don't encode your electronic mail, use regular (snail mail) for the really private stuff.

E-mail tips

E-mail is great for short messages. But for long or complex ideas, you're better off using the phone. Dr. Franklin Becker, director of the Cornell University International Workplace Studies program, comments that, "communicating subtleties through e-mail is like trying to pull an anchor line through the eye of a needle."

To improve your e-mailability, follow these suggestions:

✔ **Reread your message before sending**
   This will give you a chance to spot errors and tighten your prose.

✔ **Don't respond to messages in anger**
   If you feel like flaming someone, take a break, shut off your computer and calm down. Once you've sent an angry message, you can't recall it. The damage is done.

✔ **Avoid sarcasm and irony**
   Sarcastic humor is difficult to convey and can easily be misconstrued.

✔ **Keep it brief**
   Edit thyself. Your correspondents will appreciate crisp, succinct messages.

✔ **Preface your message with a subject line**
   When you answer a message, copy the original message (or the gist of it) so that your recipient knows what you are writing about.

✔ **Don't use ALL CAPS**
   Uppercase denotes SHOUTING in the online world and should be avoided unless you want to EMPHASIZE just a word here and there.

✔ **Save time by using an address book**
   Keep frequent e-mail addresses in your online address book. Whenever someone sends you a message, copy the sender's online address to your electronic address book.

✔ **Set up distribution lists**
   If you regularly correspond with a particular group of peo-

ple, create a distribution list and save it in your address book.

✔ **Don't send junk e-mail**
Because e-mail is so inexpensive and easy to use, some people have been tempted to send out unsolicited ads (junk e-mail) to hundreds or thousands of recipients. Don't do it. You may be thrown off the network and, if enough people are mad at you, you may be bombarded with prank phone calls, junk faxes and other miseries.

Transferring files
Most e-mail services provide a means for attaching files to messages or sending files directly to someone's e-mail address. I wish I could say that this is always easy but that would be a lie. Sometimes file transfers go smoothly, other times, everything breaks down. And it usually is not your fault.

Compressing/decompressing a file
Many online services and bulletin boards *compress* files to make them smaller so that it takes less time to transfer. Oftentimes, these compressed files are called *zipped* files. To read the file after you download it, you must be able to *decompress* or *unzip* the file. This is done using programs such as PKZIP or StuffIt. You can find these or similar programs on almost all bulletin boards or services.

Downloading/uploading a file
Most communications programs offer you a few choices when you are sending (downloading) or receiving (uploading) a file. If you want to up- or download pictures, you'll want to specify *gif* (graphics) format. For files containing words, you can specify either a *text* file (also called ASCII) or a *binary* file. If the file is absolutely straight text with no formatting, text is okay. But, if your file contains formatting (indents, underlines, bolding, changes of font or font size), you'll want to specify a binary file. This way, the file will contain your formatting commands.

If all else fails, ask your correspondent to copy the file into the body of an e-mail message and send it. You'll lose formatting but you'll get the meat of the message.

What about viruses?

A computer virus is a software program designed to disrupt computer processing. You may recall reading about the Michelangelo virus that, back in 1992, was rumored to shut down all computers throughout the world on Michelangelo's birthday (March 6th) and destroy all data. Fortunately, the rumor alerted most PC and mainframe users, and an unprecedented number of virus detection software programs were sold and installed. As a result of such vigilance, less than one percent of all computer users were affected.

According to the Antivirus Researcher's Report, 150 to 200 new viruses appear each month. Your computer could get infected with a virus if you run a program that is infected. You can get infected files from all sorts of places: programs downloaded from bulletin boards, the Internet, and other online sources, files you've copied from friend's floppy or hard disk, or even from brand-new commercial software. You can also get a virus as an e-mail attachment. Though most viruses can only infect your machine if you run a program containing the virus, there is one exception: a group of viruses commonly named the Microsoft Concept virus. These actually attack if you open any file that contains the hidden virus. Nasty.

Your first step in protection is to get a virus protection program and use it. Among the best are programs from Symantec, Central Point, and McAfee.

Remember, the program you use is only effective recognizing *known* viruses. You'll need to keep your program current by downloading new search strings (found on most online services—use the keyword VIRUS) that help you detect new viruses and subscribing to regular program updates so that you can safely remove infected programs.

How to recognize symptoms of infection

There are all kinds of viruses, and some can remain hidden in your system for months or even years. Each virus exhibits different symptoms. Here are some of the most common:

• Your system date and time stamp changes all by itself

• You notice that your programs keep growing in size

- Your system is crashing more often
- Your computer seems to be slowing down
- You see weird or comical error messages such as "feed me" or "April Fools!"
- Your disk is wiped clean

How to avoid infection
Install and use an antivirus program to scan your hard disk and each floppy disk before you copy files to your computer or run the programs they contain.

✔ Keep your antivirus program up-to-date by subscribing to the program's update service so you can receive new search-and-repair routines.

✔ Don't download programs (or files with attached programs) from bulletin boards or websites not well known to you.

✔ Don't buy or use pirated software.

✔ Avoid using software if the package is damaged or shows signs of tampering.

✔ Regularly back up your hard disk.

✔ If someone you don't know sends you an e-mail message with a file attached, don't download or open the attached file.

The Internet
The Internet is really a worldwide network of networks. It began in 1969 as a communications system created by the U.S. Department of Defense to link computers at universities engaged in defense research. Because it was designed to survive nuclear attack, the network has no central administration. In 1981, there were 213 computers linked to the Internet. Today, there are an estimated ten million host computers, 95,000 networks, and perhaps 50 million users worldwide in over 200 countries. The network now includes universities, corporations, government agencies, businesses, and even elementary and high schools.

On the Internet, you can correspond with a business-person in Taiwan, download a list of commercial fisheries from New Zealand, visit a newsgroup to discuss the latest developments in marketing, access research data located on a mainframe in Cambridge, England, listen to a realtime radio show, and play a business simulation game with participants from around the world. You can even hold a telephone conversation (see *Internet calling* in Chapter 3 for details) or participate in a video meeting.

The Internet provides a wide variety of services. We'll limit our discussion to a few that are most useful to businesses:

- Newsgroups
- Mailing lists
- Mailbots
- World Wide Web

Newsgroups

Newsgroups are loosely organized discussion groups, somewhat akin to an online bulletin board or forum. Newsgroups are organized by type: *biz* (business orientation), *rec* (recreation), *comp* (computers), *alt* (alternative), *sci* (science), *talk* (long diatribes) and *news*. To further differentiate the topic, clues are added to help you find the information you want. For example, *comp.dcom.fax* has the latest news about fax technology, *sci.med* discusses medical products and regulations and *biz.jobs.offered* posts help-wanted ads.

There are approximately 20,000 newsgroups, sending out over 30,000 articles a day—the number varies daily—on all sorts of topics. Sign up for ones that interest you, listen in for a while, and then start actively participating. In your messages, you can include information about your business, suggesting that people can e-mail you for details. Just avoid coming across as entirely commercial—the goal is to establish relationships, not shove products.

Signing up for a newsgroup is simple—just follow the directions for your particular software. How do you know which groups to join? Pick up one of the many Internet guidebooks and browse for interesting possibilities. You can also search for newsgroups using a directory or search engine (more about this later).

Anyone can start a newsgroup by sending a Request for Discussion (RFD) to *news.announce.newsgroups* and to other relevant newsgroups. If enough people "drop in" and use your group, you're official.

Mailing lists

A mailing list is just a group of e-mail addresses of people interested in a common subject. When you send a message to the list, it gets sent to the entire list. Some lists are moderated; most are not. To subscribe to a list, just send a message to the list manager's Internet address asking for a subscription. Then you'll automatically receive messages sent to the group in your own mail box. Hang on to the manager's address, just in case you want to cancel your subscription.

How do you find mailing lists? You can use a search engine and search for key words such as *business* and *mailing list*. Indiana University manages a searchable database of mailing lists. Take a look at *www.liszt.com*. Last time I looked, there were 71,618 listings.

Mailbots

You can use the Internet to send out automatic responses to incoming e-mail messages. This is a great way to send frequently-requested information such as product literature, research studies or sample articles from your newsletter. Called autoresponders, autobots or mailbots, this service is usually quoted by an Internet service provider at anywhere from $10 to $25 a month for each mailbot. For that fee, your mailbot can fire off the same response to an unlimited number of requests.

To set up mailbot service, you'll need an Internet account and a separate e-mail address for each canned response you want to send out. The system will send the message every time it receives an e-mail message addressed to its mailbox.

If you receive lots of e-mail asking for your price list, a mailbot could instantly send the list. And, if you also have frequent questions about how to join your organization, another mailbot could ship out a membership application.

Mailbots save time answering repetitive questions
Marcia Yudkin, author of *Marketing Online* (Plume/ Penguin Books, New York, 1995) set up several mailbots to handle inquiries. "I use one mailbot as an electronic brochure," Yudkin notes, "and two others for Frequently Asked Question files on freelance writing and small-business publicity." To sample mailbots, send Yudkin a message at *info@yudkin.com*. You'll get an automatic message in return.

An additional advantage of mailbots is that you capture the e-mail address of everyone who writes to you. This could be useful for developing an electronic mailing list.

The Web
If you want to see color graphics, sound, video and other goodies on the Net, you'll want to access the World Wide Web. The Web contains millions of hyper-linked electronic documents. Searching through the Web is like having a personalized, scrollable, interactive book. You just jump from link to link by clicking on buttons or underlined (highlighted) words.

Documents on the Web are found by pointing to an URL (Uniform Resource Locator). You point to it by either typing it in or clicking on it. For example, the URL for AT&T's toll free telephone directory is:

http://www.tollfree.att.net/dir800/

Once you've typed in the address, you retrieve the *home page* which lets you roam around and explore the information available in that area.

Finding information
The Internet's lack of conscious design creates a free-wheeling, do-it-yourself atmosphere. Because the Internet grew haphazardly, it is unorganized and confusing. Fortunately, there are many services that can help you find your way around the Net. You could use a directory (such as Yahoo or

LookSmart). Directories are organized by topic. You select a topic, browse through the titles, click to go to the next level down, until you find something that interests you.

Better yet, use a search engine, such as AltaVista or InfoSeek. These provide forms where you type in key search terms. The engine provides you with a list of sites that match your criteria. Typing a word like "patent" brought up 114,285 listings including articles about recent patent law cases, an example of a patent application (including electronic blueprints), a cartoon about the patenting process, and a video clip from a film on the history of the patent office. It could also bring you tons of junk—including the home page of Emily Patent, Wacky Patents of the Month, and loads of useless, poorly-informed opinions. Your search will be improved the more you narrow your cast.

Doing business on the Web
Businesses are flocking to the Web, where they see opportunities to market and display their wares. The ability to demonstrate your products online to millions of potential customers is attractive. Some of the ways you could use the Web to increase business:

- **Offer new services**
  A music store sells audio compact disks online for home delivery. Customers download a sample of a disk, and, if they like what they hear, order the entire CD.

- **Increase accessibility**
  A restaurant lists its weekly menu on the Web, provide free recipes, show a picture of the restaurant's interior, and a road map showing how to get there.

- **Expand your market**
  Specialty food companies load a Web site with photos and information about their gourmet foods. One coffee exporter lists its home page with several cybermalls, and reports over $15,000 a month in orders from Internet customers.

- **Position your business**
  Writers and artists are creating web sites to showcase their

work, pointing to other sites on the Net where their writing samples reside.

- **Generate leads**
  A car dealer hands out his e-mail address to car shoppers. Prospective customers send requests listing their requirements and receive in return price estimates, option lists, car models and colors available and an electronic order blank.

Steps for building your own Web site

If you decide to get in on the action, plan to start small. You could rent space in a cybermall to test the waters or create own site. Should you choose to go it alone, you'll need a *domain name*. This will form part of the address where customers will find you on the Internet. To sign up, call registration services at 703-742-4777 or send an e-message to *postmaster@rs.internic.net*.

Next, you'll need your own home page. For a simple site, you may be able to do the programming yourself, with the aid of an HTML editor for coding the document. Two good ones are Adobe's PageMill and Quarterdeck's WebAuthor. You'll also need software for uploading and maintaining your Web page information such as a File Transfer Protocol (FTP) utility.

A key component of Internet marketing is information. Your Web site should attract potential customers to visit, and hopefully, to return. If they find nothing but a bunch of press releases and advertising copy, they won't come back. Spend time developing useful content that answers the needs of your customers. Booksellers could tell stories about featured authors; travel agents could include packing tips and details on destinations; graphic artists could provide layout advice.

Make it easy for your visitors to leave comments by providing a guestbook, a survey, or a comments form. To capture lots of e-mail addresses, offer something free, such as a sample of your electronic newsletter. And be sure to put your e-mail address on every page.

You'll need to select an Internet hosting service that will provide the tools needed for your site. You may want a secure ordering form, or cybercash (a kind of e-money). You may opt for a shopping cart service (where customers pick items from a list and an automatic order is created for them) or the ability for customers to download software or files from your site. These services cost extra and the prices can vary remarkably. And, unless you're a programmer, you'll want to get expert help setting them up. According to Larry Newman, a California-based Internet consultant, "This industry is strictly buyer beware. You could easily pay for a Cadillac and end up with a Volkswagen." So be sure to shop around.

Once your site is up, you'll need to get the word out so people will know how to find you. Put your Web address on your business stationery, network online and write articles to post on newsgroups. Send announcements to Web-indexing services such as Starting Point and Yahoo. Write to other webmasters asking if they would be willing to link to your site. Hopefully, if you build it, they will come.

My Web site *(www.infographex.com/langhoff)*
When planning my Web site, I made a list of what I thought visitors would be most interested in and designed the pages to reflect those interests. My site focuses on telecommuting, a hot topic and the subject of one of my recent books. I created separate pages to answer frequently asked questions (FAQs), provide advice for wannabes, list resources, link to my publisher (for people wanting to order a book or look at the other titles on the list), display an excerpt from one of my books, link to reviews, even show a funny cartoon.

The site was designed by Annie Kook, a professional graphic artist, who is hosting the site for now. It attracts about 1,000 visitors a month and has brought in requests for interviews, book orders, and e-mail messages from around the world. All for only $20 a month!

Getting connected
To get on the Internet, you need an access account, which you can get via an online service, through a telephone company, or directly from an Internet access provider. Prices vary depending on the number of hours used, your modem speed

and the type of connection used. You'll pay a premium for faster services and ISDN connections. In some communities you can get free access, via a *Free-Net*. Write the National Public Telecomputing Network *(info@nptn.org)* for a list of community Free-Nets.

For a current list of Internet providers, contact The InterNIC (Internet Network Information Center (800-444-4345) or look in the Yellow Pages under *Computers-Online Services and Internet.*

You'll also need special software designed to navigate the Net and a file browser such as Netscape's Navigator (800-NET-SITE), or Microsoft's Internet Explorer (206-882-8080).

Saving time
Internet traffic jams are becoming more common. As response time gets longer, these "brownouts" cause gridlock and frustration. To avoid peak traffic hours, stay offline during the 4 p.m. to 9 p.m. time slot. Or try an offline Web delivery service such as FreeLoader, WebWhacker or Milktruck Delivery. These programs work somewhat like a television VCR, allowing people to browse Web site content offline. You select the pages you want to monitor, the system automatically logs on, checks for new content in the sites specified, and delivers the updated material to your hard drive.

A word about BBSs
Estimates vary, but there are at least 14,000 BBSs (Bulletin Board Systems) in the United States alone. Most of these systems are organized around a particular interest area or topic. There are special bulletin boards for AA meetings, police and fire safety, sex talk, fresh farm produce locators, hobbyists, home school information, job listings and employment counseling, real estate listings, mystery writers, and on and on.

Most BBSs are free; others charge a nominal fee. You can't get to a BBS from your commercial online service, you must dial in directly. BBSs are usually accessed through a local telephone number (they rarely have 800 numbers) and often have a local or community-wide focus.

To access a bulletin board, you'll need communications software such as Crosstalk or WinComm Pro. With one of these packages, you can easily dial any BBS from your PC or Mac. Getting your hands on BBS numbers is the tough part. A useful resource is *Boardwalk Magazine*, 8500 West Bowles Ave. Suite 210 Littleton, CO 80123, telephone: 303-973-6038.

Choices! choices!

If you've never ventured online, I'd suggest that you begin with one of the major online services to get a feel for using a modem, using e-mail and meeting people electronically. Such an approach is easier than trying to navigate the Internet and will help you decide whether you need to venture further. When you're ready to explore the Internet, start out by accessing it from your online service. And, if you find you're becoming a frequent user, sign up for a flat-fee service—you'll save money.

Resources

Books

The Internet Business Companion
by David Angell and Brent Heslop
Addison-Wesley Publishing Company, 1995

Doing More Business on the Internet
by Mary Cronin
Van Nostrand Reinhold, 1995

The New Internet Navigator
by Paul Gilster
John Wiley & Sons, 1995

12...Modems

· ·

Now that you've read all about all the neat things you can do online, and the business opportunities available, you're probably eager to get going. But first, you need to deal with a few technical details. To get online, you need a modem, a phone line and communications software. This chapter will help you get it all together. And, because modems are a bit tricky to set up, I've included lots of troubleshooting help.

Modem communications can help you do more than just get online. You can use your modem to streamline operations, speed workflow, and reduce courier expenses. Robin White, a California-based interior designer, uses his modem to send CAD designs to clients, who then mark them up and modem them back. White estimates that the time saved for one project was over 20 hours, resulting in cost savings of about $2,000 in design costs.

How modems work

Basically, a modem takes the digital stuff you're sending and translates it into analog format. Once the information is in analog format, it can be sent over normal phone lines. Another modem sits at the other end of the connection and translates the analog information back into digital format so it can be read or stored on a computer.

Unlike most computer equipment, the same external modem can work with either a PC or a Mac. All you need is the correct connecting cord.

Choosing a modem

To select the modem you'll need, you must determine if you want an internal or external modem, plan how you want to use modem communications (data, fax, voice or a combination), and identify what kind of line you will use (analog, digital, ISDN, cable). Finally, you'll need to pick a modem speed.

Modem location

Your first decision is whether you want the modem to be installed inside your computer (internal modem) or attached to the outside your computer (external modem).

Internal modem

An internal modem is a printed circuit board that takes up an expansion slot inside your computer. You have to take the cover off your computer to install an internal modem. This is not hard, but it's not for the fumble-fingered.

External modem

An external modem is usually about the size of a quality paperback and connects to your PC with a cable. External modems usually cost a bit more than internal ones, but are handier if you want to move your modem around. External modems have display lights that provide useful visual feedback about the status of your communication.

Credit-card sized external modem

If you have a notebook or handheld computer, you can get a credit-card sized modem that fits in the PC slot that most portable computers have today. These tiny cards perform just as well as full-sized modems and are less hefty to lug around. There's even a wireless version.

Type of connection

If you get an external modem, you'll need to plug it in your computer using either a 9-pin or 25-pin connector that plugs into the serial port on the back of your computer.

If your modem has a 9-pin connector and your computer has a 25-pin connector (or vice versa), you can get an adapter at any computer store to make them fit.

If you have an Apple computer, it's even easier to find the serial port because it's marked with a phone symbol. Actually, both phone and printer ports operate the same, and can be used interchangeably.

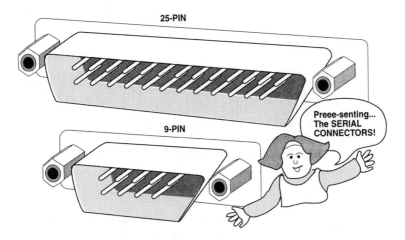

Type of modem
Be sure to get a modem that is designed to perform the task(s) you want.

Data modem
The simplest type of modem is one designed to work with *data only*. With a data modem, you can access online services and bulletin boards, upload and download files (including files containing graphics and sound bites), or log onto another computer and conduct a remote session. About the only thing you can't do is send or receive faxes.

Fax/data modem
For a few dollars more, you can get a data/fax modem. These modems have all the capabilities of a data modem plus you can send and receive faxes. To add fax capability, you must have fax software installed in your computer.

For more information on fax communications software, see Chapter 13.

Voice/fax/data modem

These specialized modems are designed to process interactive voice communications, such as audiotext, voice mail, and fax-on-demand. Some voice modems can operate as computerized answering machines. For more information, see *Voice mail in your computer* in Chapter 8 and *Fax-on-demand* in Chapter 13.

Wireless modem

In addition to standard wired modems, you can also find wireless types. Many work with cellular or PCS phones; others operate with radio signals. These specialized modems send and receives data over the airwaves. Great if you need to communicate in areas where phone lines are hard to find or are mostly digital. See *Two-way Radio Mail* in Chapter 10.

ISDN modem

If you have an ISDN phone line in your office, you'll need an ISDN-capable modem. Alternatively, get an adapter that allows you to attach an analog modem to your ISDN line. If you do this, however, you won't be able to take advantage of the higher speeds available with ISDN service. For more information on ISDN, see Chapter 4.

Hybrid modem

If you use a variety of phone lines, get a hybrid such as IBM's WaveRunner, Hayes' ISDN PC Adapter or Motorola's TA200. These modems are so smart that they can sense what kind of line you're using and adjust to analog or digital transmission, as needed. Of course, you pay a premium for all that intelligence.

Cable modem

A cable modem transmits data over coaxial cable (the same stuff that carries TV signals). The modem links to the computer using the Ethernet connection on the back of your PC.

Transmission speeds

Modems and the connecting phone lines are the slowest parts of the system, therefore it makes sense to get the speediest modem you can get. Modems are rated on the basis of their transmission speed, which is measured in bits per second, or bps. The most common speeds are 2400 and 9600 bps; and 14.4 and 28.8 Kbps (kilobits per second; 1 kilobit = 1000 bits per second). There's also modems that operate at 33.6 Kbps and 56 Kbps. Often, you'll see speeds listed as baud rate (e.g., 2400 baud).

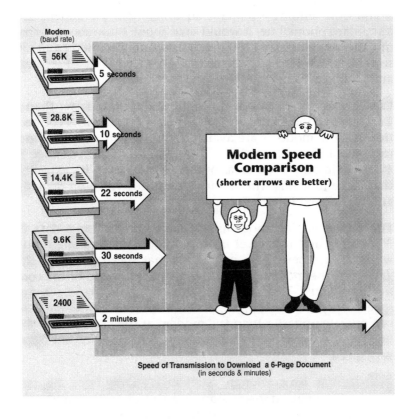

Speed of Transmission to Download a 6-Page Document
(in seconds & minutes)

A new 56 Kbps modem standard is developing. While there are some 56 Kbps modems on the market as of this writing, there are two different, incompatible standards: one operating on the Rockwell chip, the other using a chipset called K56, developed by US Robotics. This means that if your Internet service provider uses a Rockwell modem, and you use a K56, they will not be able to communicate with each other. If you purchase the wrong one, you could end up with a model that loses out in the standards war. In addition the 56 Kbps modems can only download at the high speed, uploads can't go faster than 33.6 Kbps. A final problem involves the quality of the phone lines these 56-speed modems operate on. Many are unable to support the high speeds advertised.

How fast is fast?

If you were sending a six-page document using a modem and a standard phone line, it would take about 4 minutes to transmit the document at 1200 baud, 2 minutes at 2400 baud, 30 seconds at 9.6 Kbps, 22 seconds at 14.4 Kbps and 10 seconds at 28.8 Kbps.

Fastest yet are cable modems—almost 700 times faster than a 14.4 analog modem; nearly 80 times faster than ISDN connections. These modems are available through subscription to a cable data service. At this writing, cable modem service is available in only a few areas of the United States. It's expected that cable modems will be most useful for video-on-demand and high-speed Internet access.

The fastest analog modems available can transmit data at 33,600 bps and receive at up to 56 Kbps. Unfortunately, you won't always be able to operate at that speed. Many of the gateways (the modems operating at the other end of the connection) you'll encounter are geared up to handle lower speeds only, often stopping at 9.6 Kbps. Since your modem can operate no faster than the modem at the other end of the line, your super-fast modem slows down to match the gateway speed.

And if that isn't enough, standard phone lines are not designed for data transmission and can be full of noise and other stuff that can slow down the transmission even further, even causing disconnects. Sometimes, I find that my 28.8 modem transmission speed will drop to 26, 24, 20 and even all the way down to 14.4 when encountering noise on the line.

Hint: If your modem disconnects suddenly, just dial again. Hopefully, your connection will stay up the next time. If all else fails, wait until late at night. That's when telephone traffic is lightest and your chances of a clean line are the best.

What about bandwidth?

Bandwidth is just another name for the speed at which data can move. If you have enough bandwidth, you can move a lot of data at very high speeds. Although most conventional analog modems top out at 28.8 Kbps, digital transmission can go much faster, especially on high-speed lines such as ISDN or fiber optics. Downloading a 700-page book would take only about 5 minutes over a 128 Kbps ISDN line. Using a standard analog modem at 28.8, that same book would take almost an hour to transmit.

Then there's fiber. Let's say you wanted to download a two-hour digitized color feature movie like *Jurassic Park*. First, you'd need at least 100 million bytes available on your hard drive. You'd also need a little over 12 days to download the movie at 9.6 Kbps. However, over fiber optics, which moves data at near light speed, you could transmit that same movie in just over 70 seconds (10,000 times faster than 9.6 baud). Now that's fast!

Standards

In addition to the transmission speed, modems have another set of numbers that indicate that the modem adheres to certain international standards and protocols. Here are some of the more common:

V.22   =   2400 baud

V.32   =   9600 baud

V.34   =   28.8 baud

V.42   =   28.8 baud with data compression

V.42 bis =   28.8 baud with data compression & error correction

If you see the word *bis* (French for encore) after the numbers, that lets you know that the modem has been designed to check for errors and compress data while sending. For example, a V.42 bis modem will compress the data by a factor of up to four, so that, in theory at least, a 9600 bps V.42 bis modem could transmit data at 38,400 bps. Of course, both the sender and receiver must have V.42 bis modems for this particular trick to work.

You'll also want to be sure that your modem is *Hayes-compatible* (sometimes called AT-compatible). This means that the modem will use commands that are familiar to other modems throughout the world. Almost every modern modem is Hayes-compatible these days.

There are also some nonstandard modems, which, if connected with a modem using the same standard, work at the published speed. Otherwise, a nonstandard modem will have to slow down to match the lower speed at the other end. Among the nonstandards are:

V.32terbo   =   Supports 19.2 modem speed

V.fc (fast class)   =   Nonstandard protocol - supports 28.8 modem speed

Communications software

In addition to a modem, you'll need communications software so that your computer will know how to dial and connect successfully to the computer at the other end. Major online services provide specialized software either free or for a very reasonable cost. Many new computers come with one or more communications programs pre-installed. If you plan to use your modem to dial local BBSs or call your office and download some files, you'll need general purpose communications software such as Crosstalk, or Symantec's WinComm Pro. You can often download Internet access software for free.

Getting connected

Hooking up an external modem is easy. You just plug the modem into a power outlet and the phone line cord into a wall jack. Then you plug the serial connector on the modem (that's the plug with 9 or 25 gold pins) to the serial port on your computer.

Connecting a Modem

Connecting a PC-card modem is also easy. You simply slide the card into the Type II PC slot with the gold contacts facing in. Be careful that it goes in straight. When you reach what feels like the contacts, press gently but firmly. Then attach your phone cord or *dongle* connector.

Daisy-chaining
If you need to plug in other devices (fax machine, phone or answering machine) at the same wall jack, you can daisy-chain them. Just plug your modem into the wall jack as your modem using the LINE port on the back of the modem. Then plug another phone line cord into the PHONE port on the back of the modem and plug in the next device. You can continue to do this up to five devices.

PC Modem Cards

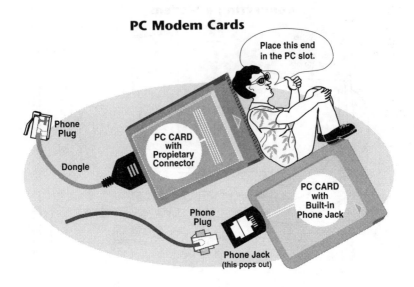

Troubleshooting

Connecting the first time

If you can't get online, don't give up. And don't think that you are stupid or something. The problem may be caused by incompatible resident programs such as virus checkers, *Ini* or *INIT* programs. You may need to modify your communications setup by adding *control strings* (special commands) to get the connection right. Try consulting your modem manual. Or, if you have an online service, call its customer service.

It took three different service assistants to help me set up America Online to work with my Hayes Accura 28.8 modem to work on a Power Macintosh. The first technician was helpful but unable to solve the problem; the second suggested a lot of control strings and finally referred the problem to their tech wizard who, after a few unsuccessful attempts, solved the problem by having me temporarily remove all the INIT files from my system folder. Somehow, these files, which control automatic functions, were preventing the modem from dialing the 800 number.

Another possibility could be a conflict over your COM port (another name for serial port). Most PCs have two COM ports these days, COM1 and COM2. (Some have four COM ports). In addition to modems, a mouse and a scanner also require COM port space. Since most communications software comes preconfigured for COM1, you may have to change the COM port to get your modem to work. There's a simple utility that lets you do this from your software; just look it up in your systems manual and follow the directions.

Protecting your modem

Modems can get hot and need air cooling to work well. Be sure that you don't block the air vents on your modem with stacks of paper or other stuff. You'll want to attach your modem to a surge protector to avoid the dangers of power spikes. Make sure the surge protector comes with a phone line connection. If you don't, a lightning strike nearby could cook your computer by traveling down the modem phone line.

Modem is slowing down

You've got a super fast 28.8 modem and you're connected to another 28.8 but for some reason, your transmission speed drops to 14.4, then 9600 and finally to 2400. What's wrong?

Nothing you can do anything about. It just means that your connection is *noisy* and the modem slowed the transmission in order to continue the connection. If you frequently encounter this problem, plan to dial at night when the phone lines are cleaner. Or get an ISDN line—digital lines are noise-free.

Modem & call waiting—an unhappy duo

If you have Call Waiting on the line, you're in for trouble when using a modem. That's because the Call Waiting tones disrupt modem communications and can cause the entire connection to be lost. You can disable Call Waiting on a call-by-call basis when you initiate a modem connection. This is usually done by dialing *70 if you have Touch-Tone service, or 1170 if you have rotary service.

However, if you are on the receiving end of a modem call, you have no way to stop Call Waiting from disrupting your connection. If you wish to keep Call Waiting service, I recommend that you get a separate phone line for modem (and fax) communications.

Reading your modem lights

External modems are equipped with status lights that let you know what is happening to your transmission. Several internal modems have software programs that will show modem status lights on your computer screen. These lights can be very helpful when troubleshooting.

A client of mine, who desires to remain anonymous ("too embarrassed"), learned how to read his modem's lights only after he was slapped with a $527 phone bill for a single call. Turns out his modem didn't hang up when the online software did. He didn't notice that the line was still active until the next time he wanted to dial out online, several days later! Learn to read your modem lights. You'll be glad you did.

Your modem may have only four lights or as many as 12. The most common are:

Modem Light Show

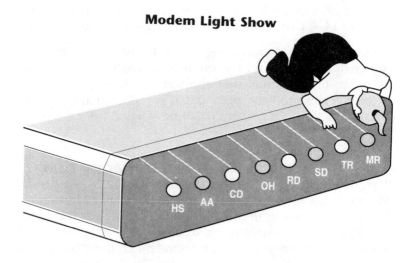

HS = **High Speed**
Indicates that the modem is operating at its highest speed.

AA = **Automatically Answer**
Lights up when you have an incoming modem call. Also, if this light is on, the modem will answer the phone line.

CD = **Carrier Detect**
Indicates that you're connected to another modem.

OH = **Off Hook**
Shows that the phone line you're using to make or receive a modem call is off-hook.

RD = **Receive Data**
Lights up when you are receiving data. You'll notice that this light flashes on and off in short bursts. This corresponds to the packets of information being sent over the line and is quite normal.

SD = **Send Data**
Blinks when you are sending data.

TR = **Terminal Ready**
Indicates that your computer is ready to make a call.

MR = **Modem Ready**
Indicates that your modem is powered on and ready to communicate with your computer.

Warning: digital line = modem killer

Some hotels, universities, and many larger office buildings have digital lines. If you travel and use a normal analog modem, you may run into them. If you're trying to connect your modem or other telephone device and it just doesn't act right, it's possible you've run into a digital line.

How can you tell if the phone line is digital? Look at the phone attached to the line. Somewhere on the back of the phone, you should see some text that indicates that it "complies with part 68, FCC Rules." You should also see a Ringer Equivalence Number or REN. If the phone has a REN number, it's analog. If no REN number appears, it's digital. Other signs of digital phones are multiple push buttons, visual displays,

built-in voice mail and keypads with special dialing or function buttons. If you're lucky, the phone will carry a warning
label saying, "Not for connection to telco lines." You can also
test the line (more on this below).

What can you do then? Find a pay phone with a modular
connector plug; pay phones are almost universally analog. Or
ask to use the fax line—fax lines are analog (unless they're
Group IV fax and those lines are quite rare). Another alternative is to carry an adapter that allows you to plug your
modem line into the handset jack. Handset lines are always
analog (otherwise, you'd be hearing the digital equivalent of
ones and zeros). You can get such an adapter from telephone
supply companies. See *Resources* for some suggestions.

Fried modem
If you suddenly find your modem not working when you plug
it into an unfamiliar wall jack, you may be in for some very
bad news. Most modems (including built-in modems and PC
modem cards) are unable to handle the higher electrical current that runs over the phone lines of digital phone systems.

If your modem encounters those lines, it burns out. To avoid
this problem, get a testing device to test the telephone port
before you attach your modem. IBM makes one called the
Modem Saver. Its a really easy test—you just insert the test
plug in the phone jack. If the light is green, go ahead; a red
light means danger. Incidentally, some of the newer modems
come with a built-in line tester.

Modem line won't release
Sometimes your modem won't hang up, even when you think
it has. This may happen if your call is interrupted by Call
Waiting or if your computer gets hung up and you have to
reset. Be sure to check your modem line to be sure that the
line has released. If it hasn't, disconnect the line from your
computer and plug it back in. That should do it.

Installing a modem on a two-line phone
If you plug your modem into the same jack as your phone,
and have a two-line jack, you may find that the modem

always dials out on line one since most modems are pre-wired to use line one.

If you want to use line one for voice calls and line two for modem calls, purchase a two-line splitter. Then plug the splitter into your wall jack, plug your phone into the receptacle labeled Line 1/Line 2 (sometimes labeled L1/L2), and plug your modem into the receptacle labeled Line 2.

Modeming internationally
There are at least 40 different types of telephone connectors used worldwide. Even within countries, you can find variations. Germany uses five different connectors; Saudi Arabia uses four.

Do some research before you travel. Find out what connectors you'll need and order them. One way to discover what you'll need is to contact your hotel. Another way is to visit Tele-Adapt's World Wide Web site. TeleAdapt specializes in international connections (*www.teleadapt.com*; 408-965-1400.)

Yet another wrinkle in international modeming are the metering pulses some countries add to their phone signals. These pulses (also called tax impulsing) are counted to determine the length of the call. They slow down modeming and can cause costly disconnects. You'll encounter tax impulsing in Germany, Switzerland, Austria, Spain, India, Belgium, Slovenia and the Czech Republic. To overcome the problem, you'll need a special filter, which you can also get from TeleAdapt.

How to dial blind
Because phone systems around the world use different signals for dial tone, busy and ringing, you may need to override your modem's configuration to connect successfully. Some modems come with a choice of country configurations. However, many designed to work in the U.S. or Canada will not be able to dial out abroad because they don't recognize the dial tone provided.

To get around this, learn how to "blind dial." This instructs your modem to ignore dial tone. Check your setup file in your modem software. You may be able to select this option. But if your software doesn't allow you to select blind dialing, you can still accomplish it. Here's how:

1. Change the initialization string in your modem's software. To do this, just add X1 to the end of the string.

2. Disconnect your modem from the phone line.

3. Instruct your modem to tone dial.

4. Listen to your modem. If you hear the series of tones your modem normally makes, you're ready to blind dial.

5. Reconnect your modem to the phone line and dial.

Although blind dialing is rather elegant, it might be easier to dial manually. Then you don't have to mess around with setup strings and pause length—you just dial as if your modem is a phone. If you don't know how to dial manually, read on.

How to dial manually

Modems will dial automatically but often it's easier to use your modem like a phone and dial out manually. Here's how:

1. Connect your modem to a phone line using a splitter or a duplex phone jack.

2. Connect a telephone to the phone line using the same duplex phone jack.

3. Be sure your modem software is configured for manually dialing.

4. Dial the desired telephone number.

5. When you hear the modem's squealing tones over the phone, instruct your software to connect.

6. Hang up the phone.

Resources

Books and magazines
Need more help? Good books to help you though the modem maze are:

How to Connect: Driver's Ed for the Information Highway
by Chris Shipley
Ziff-Davis Press, 1993
A well-illustrated, easy-to-follow book on how to connect your modem and get launched with bulletin board and online systems.

Modems Made Easy, Second Edition
by David Hakala
Osborne McGraw Hill, 1995
Written by the editor of *Boardwatch Magazine*, this books has information on buying, setting up, fine-tuning and trouble-shooting your modem.

Don't Panic! It's Only a Modem
by Esther Schindle
New Riders Publishing, 1994

Connection supplies
IBM (Modem Saver: 800-388-7080)

Unlimited Systems (Konnexx couplers: 800-275-6354)

Spectrum Information Technologies (cellular phone adapters: 214-630-9825)

Computer Products Plus (Road Warrior Toolkit: 800-274-4277)

Hello Direct (wide assortment of telephone supplies and connectors: 800-HI-HELLO)

Teleadapt (specializes in foreign connections: 408-965-1400)

13...Fax

●●

Fax is everywhere. In Chicago, prospective cooks fax in their registrations for cooking classes; in San Francisco, a talk show guest booking agent gets late-breaking industry news over a daily fax feed; in Stamford, Connecticut, commuters fax orders for home-cooked meals, which will be waiting for them at the railway station at the end of their day.

Savvy politicians are setting up fax hotlines to keep in touch with their constituents; physicians are calling fax-on-demand numbers to get the latest treatment information; libraries are faxing database search output to researchers; lawyers are faxing contracts back and forth; artists, architects and engineers fax designs for client approval. "What's your fax number?" is becoming as common a question as "Are you free for lunch?"

One out of every four phone calls made in the United States is a fax call. Over 40 million stand-alone fax machines exist in the world and the number continues to grow. The assumption today is that every business is reachable by fax. And, if you don't have fax capability, you may be losing business. As a result, many small and solo businesses are adding fax in order to present a professional image and to stay competitive.

This chapter will cover fax—including types of fax machines, fax features, fax modems and software, fax-on-demand, line sharing, and some tips on using fax more effectively.

How fax works

Fax transmission seems like a minor miracle because, with a few keystrokes, you can send an image over phone lines to practically anywhere on earth. With a stand-alone fax machine, you insert a document, key in a telephone number, and press the SEND key. The fax machine scans the document, turning the image into a patterned grid of tiny dots. These dots are then coded into digital signals. The fax machine dials the number you programmed.

It takes about 20 seconds at the beginning of a fax transmission while the sending and receiving machines perform electronic "hand shaking." They do this by sending warbling tones that check for compatibility, speed of transmission and other items before the first page is sent down the phone line. The receiving fax machine reconverts the digital signals into something resembling the original pattern. The receiving machine may keep the document in memory, on a disk or print it out.

Fax technology is divided into two major types:

*   Stand-alone fax machines - most common are thermal fax and plain paper fax

*   Fax modems (or credit-card sized PC cards) installed in your computer that, combined with fax software and your printer, perform most faxing duties

Thermal fax

A thermal fax machine uses heat to etch an image directly onto special heat-sensitive paper. This paper comes on special rolls that are 98, 164, or 328 feet long and usually 8.5 inches wide. If your fax machine has a cutter, it will cut the transmission into individual pages. Otherwise, you'll receive a long scroll and must cut it apart yourself.

Thermal fax output has several negatives. The paper is slick and difficult to write on, it tends to curl up, the image fades if left in the light, and because the paper is so flimsy, it tends to get wrinkled and torn if handled much. It's not easy to tell when you're running low on paper—until the copies appear with pink edges. This may be too late if you're receiving a long fax. Also, some marking pens interact strangely with thermal paper. You may find that a yellow highlight will, over time, turn into a dark greenish-gray smear, hiding the very words you wanted to emphasize.

Thermal fax paper capacity

Size of roll/feet	Size of roll/meters	Approx. # pages
98 feet	30 meters	100 pages
163 feet	50 meters	175 pages
327 feet	100 meters	350 pages

On the plus side, thermal fax machines are cheaper than plain paper fax. Also, they have a much smaller footprint than plain paper fax machines—important if desktop real estate is scarce. If you file your faxes, you'll find that they really don't fade all that quickly. A check in my files shows that all the faxes I've kept over the years (I couldn't find any older than 1983) are still completely readable.

Fax manufacturers are also working out the kinks (literally) in paper curling by adding an anti-curl feature that reverse-curls the paper before output. Incidentally, the larger paper rolls tend to curl less than the smaller ones and you have the advantage of not having to replace the paper so often.

Thermal fax is easy to maintain. There are no ribbons, ink cartridges or toner to replace. All you need to do is replace a paper roll, a relatively simple process. Operating costs average out to about 6¢ per page.

Plain paper fax

Instead of special paper rolls, a plain paper fax machine uses standard 8 1/2" x 11" paper, the same kind you use for photocopiers and laser printers. You can even use recycled paper. It's easier to keep track of your paper supply because you don't have to worry if you're coming to the end of the roll. Often plain paper fax machines come with more advanced features, such as broadcasting capability and memory functions.

Plain paper fax machines cost about double that of thermal fax machines. Their footprints are also larger—often about the size of a small copier. Costs per page range from 5¢ to 12¢.

Types of plain paper fax machines
Plain paper fax machines vary in their printing method.

Thermal transfer
The image is etched onto a heat-sensitive ribbon, which then transfers the image onto plain paper. Some of the older models require that you buy special rolls of rather costly plain paper, hardly an advantage.

Inkjet
Based on the same technology as ink-jet printers, this printing method provides near-laser quality for about one-third the cost. Inkjet printing is about twice as slow as laser printing.

Laser
The top-of-the-line are laser fax machines. They are fast, reliable and have the cleanest, sharpest print. However, the print output is not as good as you would see on a laser printer connected to your computer. This is because the quality of the print is determined by the image quality of the scanned fax image from the sending fax machine.

Stand-alone fax features
You won't find all these features in every fax machine nor would you want them all. But here is a list of some of the more common and useful features available today.

Activity log
A report listing documents sent or received with telephone numbers, times and dates. Useful for record-keeping and cost control.

Automatic cover page
You program your fax machine with your company name, logo, phone number and the machine automatically cre-

ates and prints a dated cover page whenever you send a document. Useful time saver.

Automatic document feeder

If your fax machine doesn't have an automatic document feeder, you'll have to stand by throughout the entire transmission and feed each sheet by hand. Unfortunately, document feeders don't always work as promised. My fax machine means well, but it often grabs two or three sheets when it should have sent only one, requiring costly and annoying retransmits. Be sure to insist on testing this feature at the store. I wish I had.

Automatic redial

This prompts the fax machine to automatically redial if it reaches a machine that is busy or the connection fails during the call.

Automatic switching

Many of the newer fax machines come with built-in switching capability, so you can use your fax line for incoming voice calls. With the aid of this switch, your fax machine can distinguish between a voice call and a fax call. These devices don't always work as well as you might hope. See *Sharing a Line* later in this chapter.

Broadcast capability

Higher-end fax machines can scan in a document to be faxed and then send the same fax to a group of fax recipients. Useful if you regularly fax to a workgroup or a specific set of clients.

Closed user groups

Lets you restrict from whom you receive faxes, and keeps your fax line open for important, expected transmissions. This is the perfect antidote to junk fax.

Copier

All fax machines can double as a copier, although the copies you get will be no better than the fax quality supported, and the output is slow. The copier function is useful for previewing faxes before sending them. If you're planning to send a complex fax, especially one with dense text or lots of illustrations, you can experiment with fine or halftone mode before actually sending the document.

Delayed transmission

Lets you program your fax machine to automatically send a fax or fax broadcast at a later time. Important if you want to send faxes late at night when phone costs are lowest.

Gray scale

If you will be regularly transmitting photographs or artwork, you'll want a fax machine with the ability to handle gray scale or halftones. For most applications, 16 tones of gray are sufficient. However, you can find machines that support 32 or even 64 shades. Gray-scale transmission is much slower than regular fax. It can take up to five minutes to send an 8 1/2 x 11 photo. If you just send text or line art, you don't need gray-scale capability.

Image enhancement

Some of the more expensive laser and inkjet fax machines can improve the quality of your incoming fax by smoothing lines and edges and defuzzing images.

Junk mail blocking

You program specific phone numbers that have been bombarding you with unwanted mail. The machine refuses to accept faxes from those numbers.

Memory

Memory has many uses. When your fax machine runs out of paper, memory allows the fax machine to save a few pages of the most recent transmission for printing out later. You can also use memory to scan in a document, and then broadcast it to several fax machines.

Out of paper reception

When your machine senses that it is out of paper, it can save a preset number of pages in memory for later printing (somewhere between 10 and 20 pages). You'll realize how useful this feature is the first time you receive a partial transmission of a much-needed document and, when you call back to get a retransmission, find that everyone is gone for the weekend.

Page cutter

This cuts transmitted faxes into individual pages rather than giving you one long scroll to cut apart yourself.

Phone

Most fax machines come with a phone, or at least a handset. You may need an attached phone if you transmit to people who don't have a dedicated line or an automatic fax switch. You'll also want a phone when calling a faxback service.

Polling

Lets you program your fax machine to automatically call a group of fax machines in sequence and receive incoming fax. You usually set your machine for delayed polling to take advantage of lower phone rates. You might use this feature if you were producing a newsletter and wanted to collect ad layouts from a group of advertisers. Of course, the fax machines at the opposite ends must be set up and waiting. Polling is also useful if you want to call your fax machine or modem and get faxes forwarded to you when you're on the road.

To prevent just anyone from calling and getting the fax sitting in the waiting machine, most fax machines offer some degree of security with their polling function.

Remote retrieval

You leave your fax (or computer equipped with a fax modem) turned on to receive and store incoming faxes. You dial in remotely and instruct the machine to send the fax to the number you specify.

Speed dialing

This works just like it does on a normal phone. You program frequently called numbers and then dial them with a one- two- or three-button code.

Status messages

Lights and/or a LED display let you know the status of your transmission, the number being dialed, etc. The display lets you check that the number you keyed in is the correct number.

Store and forward

Lets you program your fax machine to receive a document, store it in memory, and then send it on to other fax machines. This feature is most useful in large organizations that need to get information from one source and rebroadcast it to branch offices.

Transmission speed

Most fax machines today are rated for 9600 bps (bits per second) or 14.4 Kbps (kilobits per second), but you may encounter older fax machines that operate at 2400 or 4800 bps. If you fax frequently, you'll need speed to keep your phone bills under control.

Buy a fast machine unless your faxes are all local calls and you don't care how long the line is tied up. Fax machines, just like modems, will slow down if they encounter noise on the phone line. Also, the transmission speed will only be as fast as the speed supported by the fax machine at the other end.

Transmission mode

The majority of fax machines let you select the quality of the resolution. Standard mode is usually 100 by 200 dots per inch (DPI). Fine mode improves resolution to about 200 by 200 DPI and doubles the transmission time. Some machines also support a super-fine mode (200 x 400 DPI), but may require that the receiving fax have a superfine setting and be the same brand.

Fax modems

If you want plain paper capability for less cost, and you already have a computer and printer, consider a fax modem solution. By connecting a fax modem to your computer, and installing fax software, you can send any document that is stored on your computer and receive virtually any type of document. This saves time, paper and hassle. Instead of printing out the document and feeding it into your fax machine, you just send it electronically from your computer.

How fax modems work

The fax modem contains a standard data modem and the telephone signal processing found in a fax machine. To transmit a fax, you use fax software that sends your computerized file through the fax modem to a regular fax machine or another fax modem.

When you receive a fax, the fax modem answers the call and converts the transmission into a computer graphic image. In order to receive faxes, you must leave your computer on and the fax receive mode enabled.

Leaving your computer on adds about 20¢ to your electricity bill per day. Wear and tear on your computer is minimal. If you're worried about security, you'll be comforted by the fact that hackers can't get into your computer over the phone lines as long as you leave the computer in fax mode.

Incoming faxes can be viewed, discarded, printed out or saved for later. Because you can look at them on the screen, you can delete the faxes you don't want to print, saving you time and paper. The faxes come in as graphics; you can't edit the words unless you have *optical character recognition* (OCR) capability (more about this later in this chapter).

Almost all fax software operates in the background, so you can still use your computer for other tasks. When you receive a fax, you'll be notified via a beep or a popup screen. You can either view the fax on your screen or print it out on your printer.

Faxes sent directly from a computer file via fax modem are much crisper than those sent from a stand-alone machine.

They don't suffer any degradation from going through the scanning process.

Another advantage of faxing with your computer is that you can archive incoming faxes and save them for editing, printing, or retransmission later. It's easy to create and access multiple phone books, too. You can schedule delayed faxing to take advantage of lower rates after hours. You can also efficiently send broadcast faxes to groups of people just by selecting a stored distribution list.

Of course, once you have a fax modem, you can use its data communications capability to go online, send e-mail, visit the Internet, network on a business forum, explore a BBS, do business electronically anywhere, anytime. See Chapter 11 for details.

Fax modems are great if you normally send documents created on your computer. However, if you need to send items that didn't start out as images on your computer—newspaper clippings, maps, order forms or photos—you'll need to add a scanner to the mix. You then must scan in the image to be sent, check it for accuracy and, if necessary, rescan it and finally send it as a fax. Very time-consuming.

Fax files gobble disk space. A single page of faxed text is around 60K and a graphics page is around 200K. Be sure that you have sufficient space on your hard drive to support faxing from your computer. You'll want a minimum of 40 megabytes of storage space just for faxes.

Note: If you plan to install an 14.4 Kbps or more internal fax modem into a DOS or Windows machine, be sure that the computer has a 16550 UART (Universal Asynchronous Receiver/Transmitter) serial port so it can properly buffer against data loss at high speeds. Most new machines come with this port. If you don't have the latest serial port, don't despair. Serial ports are cheap. You can get a 2-port 16550 UART card for around $30.

Fax modem advantages

- Saves paper two ways: You don't have to print every fax you receive and you don't have to print your document before sending it
- Saves time—fax sending is quick and easy
- Cleaner, crisper faxes
- You can save faxes for editing, printing or sending on to someone else
- Broadcast faxing is easier

Fax modem disadvantages

- Must leave computer on to receive fax
- Can't easily fax non-computer-generated documents
- Requires lots of disk space
- Awkward to implement in a large office setting. Do you want to equip every employee with his or her own fax modem? Could be costly.

Fax software

Fax modems come with bundled software. If you've purchased a new computer lately, most likely it came with built-in fax capability and a "lite" version of fax software. You aren't restricted to the software that came with your modem, however. You can use almost any fax software with a standard fax modem. So if you want additional features (such as a choice of customizable cover sheets, OCR, or fax management capabilities), you might want to purchase a more powerful fax program.

Make sure that the software supports both incoming and outgoing fax. Some of the low-cost versions only allow you to send.

Some fax software packages come with optical character recognition (OCR) capability. OCR software converts incoming faxes into text files so you can edit the file using your word processing program. Using OCR, an attorney could receive a faxed contract, convert it to text, port it into a word processor, mark up the file with a strikethrough or highlighting font, add new text, and fax back the changes. OCR could be useful for any business that uses fax for group writing projects, editorial changes, or publishing.

In addition to text editing capabilities, OCR has the added advantage of significantly reducing the disk space required to store a fax. So far however, OCR has not lived up to expectations, and most output has errors which require careful proofreading. When using OCR, be sure to coordinate with the sender. The sender will need to send clean, non-photocopied documents with no illustrations, underlines, fancy typeface or handwritten notes. Text should be in a san-serif typeface (such as Geneva or Helvetica) and be 12 points or larger. This is a lot of bother, if you ask me. I'd rather ask the sender to e-mail the file or mail me a diskette.

Fax modem speed

Fax modems come in a variety of speeds—from pokey 2400 baud models to 14.4 Kbps. Although data modems can handle speeds up to 56 Kbps, fax cannot. If you want really fast fax transmission, you'll need to upgrade to an ISDN line and a Group 4 fax machine. See Chapter 4 for more information.

The slower fax modems cost less than the faster ones but, since the cost of faxing is directly proportional to the phone costs, you'll save money in the long run with a faster fax speed. The transmission speed will only be as fast as the speed supported at the other end. Still, it makes sense to buy the higher speed fax modem if you fax a lot and especially if you do a high proportion of long-distance faxing. For more information on selecting a modem, see Chapter 12.

Sending a fax using a fax modem

Faxing from your computer is easy once you understand that your PC is being tricked into thinking that it is printing:

1. Start up your word processing software.

2. Find the file you want to send.

3. Select Print File.

4. Change your printer settings so that your fax software is the default printer. If you have a Mac, you do this with the

Chooser. If you have a PC, you do this though the print command.

5. Click OK or hit Enter to indicate that you're ready to print.

6. A dialog box will pop onto the screen asking you to type the name and fax number of the person to whom you are faxing. Answer the questions and select OK.

7. The software will convert the file into a fax and then dial out. Most software will give you a report indicating success or failure.

Is a fax modem right for you?

To determine if a fax modem is appropriate for you, answer the questions below. A score of six or more indicates that PC or Mac faxing may be the right choice for you.

❑ Do you already have a computer and a printer?

❑ Are most of the faxes you send generated from your personal computer?

❑ Are you the only person faxing information in your organization?

❑ If more than one person in your organization normally sends and receives faxes, are you willing to install a fax board and software on each person's computer?

❑ Do you wish to keep your faxes private from others in your organization?

❑ If you wish to send non-computer-generated documents, do you have, or plan to get, a scanning device?

❑ Do you need to keep archived electronic copies of your incoming faxes?

❑ Do you need the ability to annotate faxes electronically and forward them to others within your organization or workgroup?

❑ Are you willing to keep your computer turned on all the time (or be unreachable by fax for certain periods of time)?

❑ Do you have sufficient room on your hard drive to accommodate fat fax files? You'll probably want at least an additional 40 megabytes for fax storage.

Portable faxing
You can take your fax capability with you in a variety of ways:

Mini fax machine
A small stand-alone fax machine that fits in your briefcase, glove compartment, or purse. Some can only send faxes so be sure you won't need to receive faxes if you use one of these models.

Cellular fax
Check to see if your cellular phone has built-in fax capability or takes a PC cellular fax modem card. For more information on cellular faxing, see Chapter 7.

Pocket fax modem
These external modems are a little larger than a cigarette pack and weigh in at less than a pound. Some require external power; others operate from batteries. Be sure to carry the correct cable to hook up your portable computer. Because these modems are so small and easy to hook up, many people use them as their office modem as well as their take-along modem.

PDA attachment
Many palmtops and Personal Digital Assistants such as the Apple Newton and the Sharp Wizard come with optional fax modem attachments. Some models can only send faxes and are unable to receive—be sure to check.

PC fax modem card
These credit-card sized devices fit in the card slot on portable, notebook and handheld computers. You'll need a

Type II PC slot plus fax software loaded on the portable computer.

Fax standards & protocols

The Committee on International Telephony and Telegraphy (CCITT) sets standards for worldwide fax transmissions. There are currently four international standards for fax transmissions:

- **Group 1** - The earliest standard for fax transmission. Group 1 machines are painfully slow—around six minutes per page. Unless you normally shop for office equipment at your local flea market, you probably won't find a Group 1 fax machine for sale.

- **Group 2** - Machines operating with Group 2 technology have speeds of up to three minutes per page. Don't get one of these either.

- **Group 3** - This is the current standard today. Depending on the modem, speeds range from 20 seconds per page to around one minute per page. You'll want to be sure to get a fax machine that is Group 3 compatible (sometimes listed as G3).

- **Group 4** - This standard is designed for ISDN phone lines. You won't need this technology unless you also have an ISDN phone line. Group 4 fax can transmit seven to nine times faster than Group 3.

If you're getting a new fax software package, make sure that it is compatible with your fax modem. There are three classifications for fax modems:

- **Class 1** - These fax modems use the CPU on your computer to perform many fax functions. Class 1 fax modems tend to slow down your computer when sending or receiving a fax.

- **Class 2** - These fax modems perform most of the fax functions without using your CPU. Therefore, they don't slow you down much.

- **CAS** - This is a special protocol used on a particular line of Intel modems.

Fax-on-demand

Many businesses and organizations are taking advantage of fax-on-demand technology. A customer or client calls an advertised number (often toll-free), listens to a menu of choices, keys in a one- to five-digit number to request a particular document and enters the fax number to which the document should be sent. Moments later, the fax is on its way. Another method requires that you call from your fax machine and the fax is sent directly to that machine. Instant customer satisfaction!

According to the authors of *Fax-on-Demand: Marketing Tool of the 90s*, if callers are given the option of receiving information from an electronic bulletin board or via fax, they select fax delivery by a margin of three to one.

Fax-on demand can save you money too. You can fax a letter from your PC in about one-tenth the time it would take you to manually print and then fax it. Fax-on-demand (FOD), also called fax-back, is used in a variety of ways:

- Mail-order companies can fax pages from their catalogs, brochures and price sheets to callers requesting information.

- Travel agents offer a menu of destinations and prospective travel dates. When the customer makes a selection, a list of restaurants, recreational activities, theaters, exhibits and business service numbers is delivered.

- Hardware and software firms can provide technical support via fax.

- Booksellers can fax-back catalog information or even the first page of a thriller to entice an order.

- Political campaign offices can provide prospective voters and contributors a menu of position papers.

- Real estate agents can set up a system that briefly describes each property. The caller makes a selection, and the system faxes a floorplan, photo and/or marketing flier.

Insider advice

Dan Poynter, of Para Publishing in Santa Barbara, California, operates a successful FOD system to promote both his line of books and his publishing consulting business. His customers can call his office at any time day or night and get the information they need. According to Poynter, "A quick response to customer requests for information can mean the difference between making a sale or losing it."

Fax-on-demand saves him money in a variety of ways. No more printing brochures or storing an inventory of documents. No more valuable staff time spent taking phone requests, typing up labels, and stuffing envelopes. No postage costs or even sales taxes. Not even any telephone charges associated with faxing documents. Here's how it works:

Poynter uses PhoneOffice voice/fax boards from Edens Technology installed on old 386 computers with 40-megabyte hard drives. The boards cost him $445 each. Two dedicated phone lines are linked to the fax boards in a daisy-chain fashion. The lines come with rollover service from the local telco so that if the first phone number is busy, the second line will pick up.

Poynter loads the documents (brochures, newsletters, news releases, price lists, etc.) onto the computer's hard drive. This can be done by faxing them in or, for higher clarity, loading them from disk. He composed a table of contents listing all the materials available and assigned a number to each document. He then planned and recorded a voice menu, prompting callers to key in the desired document numbers or requesting a list of the available documents.

When a customer calls Para's system, from his or her fax machine, (s)he hears the voice menu, punches in the desired document numbers and presses the START button on the fax machine. The system then automatically sends the documents requested. Everything takes place during the same call and is paid for by the caller. Slick!

But Para's system doesn't stop there. Because Poynter's publishing know-how is in great demand, he offers a selection of *Instant Reports*, ranging in price from $1.95 to $9.95. These reports provide useful tips on such subjects as getting on radio and TV talk shows, resources for the cookbook writer, and how to locate the best printer. Callers key in the desired document numbers, and their fax prints them out, along with an invoice. As Poynter says, "We're making money, even while we sleep."

Operating costs are almost nil. Poynter pays only for two standard analog residential phone lines (around $24 a month for two), an additional $2.50 for rollover service and the electricity to run two computers all the time.

Setting up a fax-on-demand system

To set up a FOD system, you'll need a dedicated computer, a dedicated phone line, a voice/fax board and software that supports fax-on-demand. You can get systems that will work on only one phone line, but many require two at a minimum. Some systems offer the ability to have the sender (you) pay for the outgoing fax call and some—mostly the more expensive systems—have the capability to take credit card orders. FOD systems range in price from $300 to over $10,000. The higher-priced systems come preinstalled on its own dedicated computer and have more features. However, if you can handle a screwdriver, you can easily install the hardware yourself. See *Resources* for a list of companies that sell fax-on-demand software.

If you're not sure that fax-on-demand will work for you, consider trying it first. Start out with a fax-back service provider who will take care of all the details for you. You'll find them listed in your local Yellow Pages under *fax transmission services*.

Calling a fax-on-demand system

Calling a fax-on-demand (FOD) system is easy if your fax machine has a built-in phone. You just dial the number, listen to the voice menu, key in your choices and press the START button—if you're connected to a single-line fax-back system (where you pay for the call). If you're calling a two-line system (where the fax machine calls you back), you can call from any touch-tone telephone and key in your fax telephone number when prompted.

You'll run into difficulty when you want to dial a single-line fax-back system using a fax modem connected to your computer (or a fax machine with no handset). The easiest solution is to plug a splitter into the wall phone jack. Then plug a telephone into one of the jacks and your fax machine or fax modem into the other. Call the FOD machine by telephone, listen to the voice choices, key in the documents you want and press the START key on your fax.

Getting connected

Hooking up a fax machine is simplicity itself. You just plug the machine into a power outlet and the phone line cord into a wall jack. Be sure to locate your fax machine in a secure area, away from the public eye, but close enough to be able to spot an incoming fax or a low paper condition.

Switch first

If you're using an external line switch, just be sure that the switch is the first device on your line (closest to the wall plug) so that all devices (phone, answering machine, fax) are driven by the switch. Otherwise, the device will be unable to switch calls properly.

Connecting A Fax

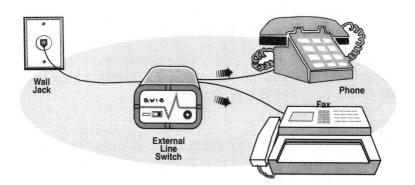

Sharing a line

Most fax machines today come with some kind of line-sharing capability. One of the simplest involves the fax machine "listening" for a specific fax tone (called CNG) sent out by most fax machines. If no tones are forthcoming, the fax machine rings your phone. This can be pretty strange for your callers however, because the line is picked up and they hear nothing for up to 20 seconds. Unless they have been clued in on how to interface with your particular system, they will most likely hang up.

Some fax machines provide a short voice instruction saying something like "If you'd like to send a fax, please press 1; otherwise stay on the line and someone will help you." Another drawback of this type of switch is that CNG tones are not generated during some types of fax calls. Many older fax machines do not emit CNG tone. Tones are not generated during manual fax calls where the person sending the fax manually dials you, waits to hear fax tones from you and then presses the send button. CNG detection is also adversely affected by noisy line conditions. No CNG tone, no fax.

Fax/answering machine compatibility

A better solution is to get a fax machine with an answering machine interface. Then you plug your answering machine into the fax machine. The fax machine sits in the background, allows your answering machine to play your greeting and listens for the incoming fax tones. If it doesn't hear anything after a few seconds, it switches the call to your answering machine. If you use this method, be sure to keep your answering machine greeting short—15 seconds or less—otherwise the sending fax machine will "lose patience" and hang up. Although you will only hear one ring before the answering machine picks up, your callers will hear three rings at least. And, with this solution, your fax machine is always going to answer your line so you won't be able to use your answering device's toll saver function.

Distinctive ring

The best choice for line sharing, if it's available in your area, is distinctive ring, a phone company feature that assigns multiple phone numbers to the same line. Each phone number rings with a different cadence so you can tell what type of call is ringing in. If your fax machine or answering device is distinctive-ring capable, it can be programmed to switch one type of ring to the fax, another to your answering machine or voice mail, etc. You could also buy a separate distinctive-ring switch that serves the same purpose. Distinctive-ring detection is highly reliable. For more details on distinctive ring, see Chapter 3.

Daisy-chaining

If you need to plug in other devices (a phone or answering machine, for example) at the same wall jack, you can daisy-chain them. Just plug your fax machine into the wall jack using the LINE port on the back of the machine. Then plug another phone line cord into the PHONE port on the back of the machine and plug in the next device. You can continue to do this with up to five devices.

Daisy Chaining

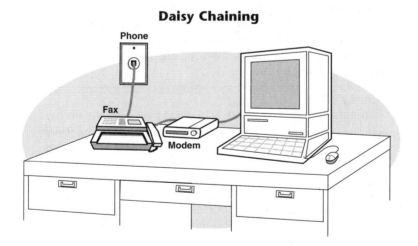

Do you need another line?

Before you run out and buy a fax switch, take a hard look at your communication needs. Though you will save some money in phone charges by using a switch on a single phone line, are you risking the possibility of losing business by sharing fax and voice calls on the same line? Fax calls often take a long time (especially if you have a slower fax modem or a noisy line). If a fax is taking up your only line to the outside world, how can you be sure that potential clients and customers aren't dialing elsewhere?

If you expect to send and receive a lot of fax calls, it's best to install a separate line. You'll appear much more professional, and your incoming and outgoing faxes will be less likely to be delayed. What's more, you'll avoid misdirected calls and irritated callers.

Fax line hunting

If you have more than one line and your lines hunt or roll over, put your fax and a fax switch on the last line of the rotary. This is less disruptive and lets you give out a dedicated (almost) fax number to your callers. For example, if you have three lines that hunt (555-1001, 555-1002 and 555-1003), publish 555-1001 as your voice line and 555-1003 as your fax line. Callers will dial 555-1001 and, if that line is busy, their call will automatically ring 1002. If 1002 is busy, their call will wake up the fax switch and it will route the call appropriately. Fax calls will always come in on 555-1003. It is less likely that the third line will be busy, thus giving your fax calls a better chance of getting in.

In my office

I originally tried to run fax and voice over one line on my home office phone, turning on the fax machine only when I expected a fax. But I soon tired of scrambling under my desk (where the fax is kept), hurrying through calls when I expected a fax, and feeling greatly stressed. So I installed a second line to handle fax and data calls. I occasionally use the second line for outgoing calls but usually keep it clear for incoming faxes. It's great—gives me peace of mind.

Connecting portable fax

Portable fax machines (other than wireless models) connect to a modular phone jack, just as desk models do. If you can't find a modular phone jack to connect to, you might want to invest in an adapter that allows you to connect to the phone's handset. Companies that make such devices include Unlimited Systems Corp. (Konnexx couplers; 800-275-6354), and Computer Products Plus (Road Warrior Toolkit; 800-274-4277).

Multifunction machines

Several manufacturers are selling combos that perform fax as well as other functions. You can get combos that come with a cordless phone, send-and-receive fax capability, and answering device, or a multifunction device that combines an ink jet printer, plain-paper fax and copier in a single package. Combo machines are especially useful if you're short of desktop space and are often less expensive than separate components. On the other hand, you're really out of luck when any one function breaks down.

Internet faxing

Don't have a fax machine or fax modem? Or the fax machine is busy? You could send your document through an Internet-based fax service. You just download the required software (often free), attach the document you want to fax to an e-mail message, type in the phone number, and that's it. You'll pay a per-page fee ranging from 10¢ to 20¢ per page for domestic faxes. Some services charge a monthly fee, whether you use them or not. The costs you pay will be offset by potential long-distance savings. Here are a couple of services to check out:

- FaxStorm *(www.netcentric.com*; 800-POP-WARE)
- Faxaway *(www.faxaway.com*; 800-906-4329)

Do you need your own fax machine?

If you only need to send or receive an occasional fax, consider a fax service bureau. You can send an e-mail message as a fax via online services such as America Online, CompuServe or an Internet fax service over a modem. Or fax from a local business or service such as Mail Boxes, etc.

Service bureaus also provide broadcast faxing. You give the bureau the document and a list of fax telephone numbers, and they do the rest. Prices are usually based on time of day. You can often get a volume discount. In addition to fax sending functions, service bureaus also provide fax mailbox services. A fax mailbox could be useful if you're out-of-town and still need to get those faxes. You forward your line to the fax mailbox service (using telco Call Forwarding), and they catch and store your faxes for you electronically. You can call in to retrieve your faxes at your convenience.

If you use the same phone line for fax and data, you might want to use a fax mailbox service when you plan a long online session without worrying about missing a fax. Just forward your calls before you log on.

Fax mailbox service costs vary depending on the number of faxes received and the services you sign up for. Most phone, pager, and wireless companies offer fax mailbox service.

Do the numbers
When analyzing your fax needs, be sure to do the numbers. Kim Ecclesine, a Bay Area multimedia writer, took the fax plunge over five years ago. Previously, she drove about a mile to the local copy shop to send and receive faxes. Kim calculates that she sends and receives about 30 faxes a month now and has saved over $16,000 in fax charges over the years. According to Kim, "I laugh every time I pass that copy shop. I spent $600 to save $16,000. What an investment!"

Send via e-mail or fax?
If your file is already on your computer, and it's mostly text, consider sending it via e-mail rather than faxing it. You'll save money.

Lets say you send a 10-page fax from San Francisco to New York during business hours. That fax would take a little over three minutes at 9600 bps and cost approximately $1.20 in phone charges. If the line was noisy and transmission dropped down to 2400 bps, the fax would take over 12 minutes and cost about $4.75. The same file, sent via e-mail on CompuServe, would cost about 2¢ plus local phone charges for the online connection. And, when the file arrived, it would be clean, readable, error free, and computer-readable.

Using e-mail to send documents also means that you don't have to mess with OCR (optical character recognition).

Use fax transmission when you want to send information that contains graphics such as blueprints, photos, and line art or handwritten notes. Use e-mail for the rest. This, of course, assumes that the recipient is accessible by e-mail. If your correspondents don't have an e-mail address, talk them into getting one. Show them the numbers. They'll be glad they switched.

Sending faxes overseas

Faxing across international borders can be difficult. You need a bewildering assortment of telephone prefix codes, the telephone tones don't always match U.S. tones, and telephone numbers can stretch for miles (or so it seems). Here are some tips to guide you through the maze.

✔ Use international codes. All international calls originating in the U.S. must be preceded by 011, followed by a country code as well as a city code. To dial telephone number 24433 in Berlin, the dialing sequence would look like this:

> 011-49-30-24433
>
> 011 = International code
> 49 = Germany
> 30 = Berlin

Hint: You'll find a list of country and city codes in the front of your telephone book.

✔ Check to see that the number you're dialing is current. Some countries are updating their dialing plans. Great Britain recently added an extra digit to the city code. Almost all city codes in the UK now begin with a 1, so you may need to add a digit. For example, London used to be 71, so now it is 171. To dial 123456 in London, the dialing sequence would look like:

> 011-44-171-123456

✔ Change the timeout. Timing can also be an issue. Some international calls take more than a minute to complete the connection. By this time, most fax machines have lost patience and hung up. If your fax software allows you to change the time before the fax gives up, do so.

✔ If you find yourself unable to get a fax through, give up and call in the experts—a fax service bureau. Or send your international fax through an online service or via the Internet.

Tips for effective faxing

Keep a sufficient supply of fax paper on hand. If you use paper rolls, visually check your supply at least once a week and before each week-end. Never allow yourself to run out.

✔ If fax is your lifeline, consider adding an inexpensive fax machine on the same line as your fax modem. Then you program the modem to pick up on the first ring and program the cheaper machine to pick up on the fourth ring. That way, if your computer is off, you'll still get the fax. A good backup.

✔ Make it easy for your clients and customers to fax you. Print your fax number on your stationary, business cards and order forms.

✔ For the cleanest fax, use a sans serif typeface such as Helvetica or Univers and a point size of 10 or more. And for faster faxing, avoid underlines and lots of black.

✔ The 3M Company sells a tiny post-it fax routing slip that you can affix to the first page of your fax. Saves time, money (one less page to transmit) and trees.

✔ Use a formal fax cover sheet when sending to large organizations or when the information contained is confidential. It should, at a minimum, contain your logo, address, voice and fax numbers, as well as the name, voice and fax number of the recipient, date, and (very important) the number of pages sent. If you want the recipient to call you to confirm receipt, add a simple check box for that, too.

✔ Create a company policy for fax use. Set guidelines for what kind of information may be faxed (urgent, short, informal communications) and when (after hours for long faxes; after 8 p.m. if an international call). To discourage personal use, make sure that staff know that you will be regularly monitoring fax usage.

✔ If you have a fax modem, you can still include your signature or logo on a computer-generated fax. You'll need to scan in the logo or signature, save it as a bitmapped graphic file and paste it into each document requiring it. If you don't have a scanner, fax yourself a copy of your letterhead from a local fax service.

✔ Most major companies fax their press releases, and you can, too. Using the broadcast feature available in higher-end fax machines and most computerized fax software, create a master list of press contacts. It should contain the editor's name and title, and the publication's name and fax phone number. Then you load or key in the press release, press the START button and you're in business.

✔ Don't list your fax number in a fax directory. To do so will probably lead to an avalanche of junk fax mail. If you do receive junk fax, call the sender and ask that you are taken off their list. The Federal Communications Commission requires that anyone sending fax ads must include name, phone number, date and time of the transmission on the first page of the fax.

✔ Be careful that you don't cross the line between junk fax and meaningful fax advertising yourself. It's not likely that you're going to get business from someone if you just tied up their fax line when they were anxiously awaiting an important incoming fax. And, although fax paper is relatively inexpensive, lots of junk fax recipients complain mightily that they got an unwanted fax and had to pay for the paper.

✔ Need to send an extra-wide fax? Architects and designers that need to fax blueprints, posters or newspaper layouts can use a two- or three-foot wide fax machine made by The WideCom Group (905-566-0180).

✔ If you need color fax capability, check out Color Communication Corporation's Color Express software (Internet: *www.color-comm.com*; telephone: 415-966-8737.)

Troubleshooting

Fax machines are simple to operate. Most problems seem to be caused by installing paper improperly. Check your manual first.

Call Waiting

Call Waiting and fax transmissions do not mix. If you have Call Waiting on the same line that you use for fax, be sure to disable it (by dialing *70 for touch-tone service or 1170 for rotary service) before you send a fax. This won't help you if Call Waiting tones interrupt an incoming fax, however. The Call Waiting tones will cause the fax modem or machine to disconnect. Your best bet is to get a separate phone line for fax communications.

Slow transmission

If you find that your fax transmission is slowing down, not to worry. The fax will automatically slow down when it encounters noisy phone lines or a slower fax on the other end of the line. The transmission speed will only be as fast as the speed supported by the fax machine at the other end of the connection.

Resources

Books and newsletters

Fantastic Fax Modems
by John A. McCormick
Windcrest/McGraw Hill, 1994
Helps you select, install and customize a fax board for either your Macintosh or PC computer. There are tips on international faxing, secure faxing and portable faxing. It also contains listings for hundreds of products.

The Fax Modem Source Book/Book and Disk
by Andrew Margolis
John Wiley & Sons, 1996
Everything you ever wanted to know about fax modems plus some. The disk contains fax source code and utilities.

Fax-on-demand systems & software

FaxFacts
Copia International
630-682-8898

PhoneOffice
Edens Technology Corp.
714-641-1235

QuadraFAX
Brooktrout
800-333-5274

FaxFriday
Bogen Communications
201-934-8500

FaxLine
Ibex Technologies
800-289-9998

Speech Master
SpeechSoft
609-466-0757

14...Telecommuting

•••

Imagine slipping out of bed a mere five minutes before you need to be at work, grabbing a cup of coffee and heading down the hall to your home office. There, you dial into your central office computer, pick up your e-mail, check your calendar, and settle down for a quiet, productive day of work. During the day, you manage to not only fulfill your work obligations, but also take the dog for a noontime walk and fit in an after-school conference with little Jenny's fourth-grade teacher.

This may sound idyllic, or impossible to realize, but it's happening all over America. Currently, over 12 million U.S. workers telecommute at least once a week, according to consultant Jack Nilles, author of *Making Telecommuting Happen*. Nilles estimates that the number of telecommuters continues to grow at a rate of about 20% a year. This is a trend worth joining.

Telecommuting brings work to people by moving jobs to workers, instead of moving workers to jobs. Work is once more something people *do* instead of someplace where people *go*. Employees work from their homes, from satellite offices and from neighborhood work centers, using virtual office technologies such as personal computers, modems, e-mail, voice mail, cellphones, pagers and fax.

Tasks that are most appropriate for telecommuting are jobs where a person often works alone, handling information such as reports, proposals, data or research. Writers, salespersons, accountants, programmers, graphic artists, researchers, engineers, architects, public relations professionals—all are prime candidates for telecommuting. But you don't have to telecommute full-time to reap the benefits of telecommuting. Since most telecommuters spend two to three days a week at their central office, it's easy to save project work, reading, report drafting, research and the like for the days at home and use office time for face-to-face meetings, team sessions, use of office equipment and the like.

You don't have to have a lot of fancy equipment to start telecommuting. I've been a teleworker for years. When I began, I just had a computer, a telephone and an answering machine. That sufficed for many years. Now, I work solo for weeks at a stretch, staying in touch with my clients using fax, phone calls, voice mail status reports and e-mail.

> **Telecommuting lawman**
> And you don't have to have a traditional deskbound job to telecommute. Take Sgt. Gary Hansen of the Los Angeles Police Department, for example. Hansen, a veteran with over 25 years of service, works at his home in Sunland/Tujunga once or twice weekly. On his work-at-home days, Hansen avoids the crowded freeway he normally traverses to his offices in downtown Los Angeles. Over the course of a year, Hansen saves over 4,400 commuting miles and the equivalent of almost three work-weeks formerly spent on the road.
>
> Hansen uses his home time to read cases and work on an operations manual. "I get nearly twice as much work done at home," he says.

This chapter deals primarily with the technology of telecommuting, including suggestions for what you need to get started as a telecommuter. I'll focus on some technologies that are particularly appropriate for telecommuting, such as remote computing, document conferencing and videoconferencing. Also included are some tips on staying connected when you're on the road or working from home.

Why telecommute?

Telecommuting makes good business sense. According to Link Resources, telecommuting can increase employee productivity by roughly 20 percent. Productivity increases because telecommuters experience less distraction at home, are able to work flexible hours, suffer less stress, and, as a result of these desirable working conditions, stay highly motivated.

Because telecommuters will work with a cold or other minor ailment that might keep them away from the office, telecommuters actually work longer hours and more workdays than the average employee. Absenteeism drops, often by half, and long-term disability costs are also reduced.

Telecommuting provides an employer with another major advantage—one that is only realized when an emergency occurs. Because telecommuters can work in a distributed fashion, often a company can keep going even if the offices are destroyed. For example, companies with telecommuting programs were able to get back in business within hours after the Northridge earthquake in 1994. And, after a fire shut down the *Dallas Times Herald*, journalists working from home were still able to get the newspaper out. Telecommuters were the only employees able to get to work when the blizzard of '96 closed offices all along the East Coast.

Passage of the Clean Air Act by the U.S. Congress in 1990 has spurred companies to find alternatives to commuting for their employees. So naturally, one of the largest beneficiaries of telecommuting is the environment. With fewer commuters on the road, traffic congestion is reduced. If we all worked at home only once a week, we'd cut traffic by 20%. Energy is conserved, demand on transportation infrastructure is reduced, and air pollution is cut significantly.

Start-up costs are low. Pacific Bell budgeted $4,000 for each work-at-home employee. JC Penney provided two phone lines (one for incoming calls; the other for outgoing) as well as a PC and some other equipment. Penneys found that start-up costs were less than half that of an equivalent office installation.

Telecommuting also expands the radius of the labor pool. For example, New York Life has several computer programmers working from home—one of them lives in Nevada. And, for those who make real-time phone contact with clients in Europe or Asia, telecommuting makes even more sense. Who wants to get to work daily at 4 a.m. just to make phone calls to customers in Finland or France?

Facility savings

Companies with telecommuting programs save big on facility costs. If workers share offices, using them on alternate days for example, the amount of floor space needed for office workers is significantly reduced. IBM, for instance, consolidated 400,000 square feet of office space into a 100,000 square-foot facility at Cranford, New Jersey.

Their facility is designed on the hotel principle. You check in with a computerized secretary who assigns you a cubicle and switches your calls to the appropriate cubby. They're using FlexiMOVE software (800-IBM4YOU). Office consolidation and *hoteling* has paid off. According to Debbie Zilau, IBM's Workforce Mobility Project Manager, "Our people are much more productive and are spending much more time with their clients."

The advertising firm, Chiat/Day, took virtual office technology one step further when it replaced each employee's office with a cordless phone and a portable computer. When a staffer needs a desk for a day, he/she is assigned a tiny one with wheels, no less. Each employee has a locker for storing personal items. The office space was redesigned to encourage group work. Sofas, conference areas and inviting open spaces replace the former office cubicles. Founder Jay Chiat explains that now employees come to work to use the office as a resource, to meet with other people, and use libraries and reference tools.

The downside

Telecommuting has its negative aspects as well. It's not for everybody—isolation, procrastination, even boredom—get to some. Temptations such as a much-too-handy refrigerator, neighbors who think work-at-homers aren't really working, household chores, and family distractions can easily undermine others.

Because their office can be anywhere they park their portable, workaholics often find it difficult to end their day. Burnout among telecommuters is a real concern as companies continue to downsize and heap more and more work on the remaining workers.

Others miss the social aspects of working with other people and networking by the water cooler to keep up-to-date. Lastly, there is the fear that they will be left out of the loop, ignored for future promotions and viewed negatively as a non-team player. This fear appears to be unfounded. Dallas-based telework consultant Joanne Pratt learned, when surveying over 17,000 telecommuters, that remote workers receive promotions at a greater rate than non-telecommuters. Could it be because their productivity is so high?

To avoid some of these negatives, many companies restrict telecommuting to no more than three times a week. Others, however, require only once-a-month onsite meetings. And the number of full-time telecommuters is on the rise as companies begin to turn their employees into virtual office workers and reap substantial real estate savings.

Survey: the successful telecommuter
Successful telecommuters are disciplined self-starters who like to work solo. Take the survey below to measure your chances for success (the more check marks, the higher your score)

❑ Are you well organized and goal oriented? At the very least, you'll want to brush up on time-management skills.

❑ Are you effective at controlling distractions? Family, neighbors, and pets will compete for your attention.

❑ Do you work well with a minimum of supervision?

❑ Are the social aspects of the office environment relatively unimportant to you?

❑ Are you an effective communicator? You'll need to be— most of your interaction will take place over phone or e-mail.

❑   Can you set aside an area of your home to be used exclusively as an office?

❑   Are you comfortable with the idea of working solo?

❑   Can you get along without office support systems and personnel? (No more copier, message-takers, typing pools. No computer-guru or network administrator at your beck and call.)

❑   Can you easily get along without in-office reference material (or arrange to get copies for home)?

How to get started

If you don't currently telecommute, and want to convince your boss that it makes sense for your organization (and of course, for you), do some research. Crunch some numbers and make a formal presentation, showing how telecommuting will save the organization money. Be sure to include cost/benefit figures.

Telecommuting Cost/Benefit Analysis	
Productivity increased	$7,500.
Absenteeism reduced	453.
Office space savings	1,000.
Parking space savings	240.
Annual total savings per employee	$9,193.
Annual cost savings based on telecommuting 2 days/week	

Assumptions: Figures are based on an annual salary of $50,000, productivity increase of 15%, absenteeism reduced by 10%, parking at $600/year reduced by 40%, and use of central office facilities of 100 square feet @ $25/square foot rent per year reduced by 40%.

Shower you manager with clippings from newspapers and magazines highlighting successful telecommuting programs. Try to find information about other companies within your own city and industry that are telecommuting. To allay your manager's fears of losing control, suggest that you start out slowly, maybe only for one day every two weeks.

Once you have initial approval, all you need to do is to organize for telecommuting. This requires a formal agreement between yourself and your organization, spelling out the goals and expectations. At the very least, the agreement should contain:

- A list of the equipment your organization will supply

- A list of the equipment you will supply

- A description of your home work space (done to satisfy Worker's Compensation requirements)

- A list of expenses your company will reimburse for (e.g., copying, telephone, supplies, etc.)

- Your normal work hours

- How your job performance will be measured

- Spell out who will be responsible for maintaining company equipment at your work space

Telecommuter's tool kit - at home

According to telecommuting guru Jack Nilles, telecommuters need the same technology that they use in the office at work. So, if you use e-mail, a phone and a computer at work, you'll need the same or an equivalent at home or on the road. Here is a list of the most common equipment mentioned by telecommuters:

Separate telephone line

This is absolutely essential if you plan to use your phone for modem or fax purposes, or if you make a high volume of voice calls. Your business callers and coworkers must be able to reach you easily. They should not be subjected to frequent busy signals. Nor should you expect your family to stay off the home line during business hours. Often, your employer will pay for your extra line.

Call forwarding

If the work site is not too far from the home office, forwarding calls from the office to home makes sense. Your office would be billed for the cost of the call forwarded from your office to home. If the distance from office to home could run up toll charges, arrange to have your office phone answered by voice mail, an answering device or receptionist. It's cheaper.

Personal computer

If you use a computer at work regularly, you'll be lost without some kind of computing capability at home. Many organizations have extra laptops on hand that you can check out for working elsewhere. The computer comes preloaded with the software needed to perform your job.

You do not have to have the same kind of computer as you do in the office to telecommute successfully. If you upload and download files via modem, you can use a Mac at home and a Windows-based machine at work (or vice-versa). One of the easiest ways to do this is to attach your file to an e-mail message.

Modem (or fax modem)

If you plan to communicate with your main office by e-mail or access files remotely, you'll need a modem. Since fax modems cost just a few dollars more, it makes sense to add this capability.

E-mail

You need to be part of your office communications system, regardless of where you are. If your office has a separate e-mail system, you may not be able to send e-mail to your organization, unless the two mail systems have a gateway. Check with your telecom administrator at work and ask if you can get modem access to your e-mail system. If a gateway exists, all you need is an e-mail address, which you can get by signing up for a public online or e-mail service. The e-mail system you use at home doesn't have to be the same system as your office's e-mail.

Voice mail or answering machine

Since most organizations use proprietary voice mail, you may not be able to link directly to its system. But you'll need a method for getting messages when you're on the phone or unavailable. Voice mail is the best choice because it can handle multiple incoming calls and take a message every time. You don't want callers to experience busy signals.

Tips for telecommuting from home

Here are a few suggestions to make working from home easier:

✔ Set up a communications routine with your office. Report in at least twice a day—once in the morning and once at the close of your day. This could be via a direct phone call, e-mail message or voice mail.

✔ Be kind to office support staff. Often their opportunities for telecommuting are limited, yet their help makes it possible for your to telecommute easily. Be lavish with thank yous. Give gifts.

✔ Need to meet with far-flung colleagues, yet avoid travel costs? Schedule a teleconference. According to Daniel Kehrer, author of *Save Your Business a Bundle,* the average teleconference meeting consists of 10 participants and lasts about 50 minutes and will cost about $170. That's only $17 per person. Compare that to the cost of business travel.

✔ If you initiate the call, your call is free. All other conferees pay a set-up fee (around $10 each) and about 50¢ a minute. In-state rates differ. Check with your long-distance operator for details.

✔ Set up a private conference with your workgroup on a BBS or online service using a "private chat room." There, you'd be limited to typed communication, but costs could be minimal because each participant could dial a local access number.

✔ Hold a videoconference. Kinko's copy centers, teamed with Sprint, now provide room-based videoconferencing at over 100 of their branches. Costs: $300 per hour for a two-way meeting; $630 per hour for a three-way meeting. Might sound expensive to you, but compare that to the costs of air fare and hotels. To set up a conference, contact Kinko's for the room (800-743-2679) and Sprint for the phone link (800-669-1235).

✔ Stay vigilant about viruses. Many computer users spread viruses without knowing it. One of the most common ways to spread infection is by passing floppy disks that contain infected programs between computers. In addition, viruses can be transmitted over phone lines, either when downloading from electronic bulletin boards or when porting programs from a remote host. Be sure to use a virus protection program, and back up files frequently.

✔ Post a calendar in a convenient spot at the workplace that shows what days you will be telecommuting and the number(s) where you may be reached.

✔ On the days you're telecommuting, be sure to answer your home-office phone in a professional manner. If you just pick up the phone and say "hello," you'll confuse your business callers. If you're unsure what to say, a simple "Good morning" or "Good afternoon," followed by your name will be fine.

✔ Use your office answering machine or voice mail to provide status to callers and change it daily. Do the same at your home office.

✔ If you don't want to provide your home office number over voice mail or answering machine, arrange for a paging service to contact you.

Telecommuter's tool kit - on the road

Portable computer

Most often listed as "the single piece of equipment I couldn't live without," a portable computer is often essential. Look for one that has a built-in pointing device (one less thing to carry; easier to use on a crowded plane). You'll also want a built-in modem or a Group II PC slot for a modem card.

Mobile phone

Be ready to conduct business anywhere—well, almost anywhere. Try to get a phone that has modular plugs for attaching a modem and headset. See Chapter 7 for the whole story.

Paging device

You'll need to keep in touch with the office and be able to get e-mail or alpha pager while on the road. Paging devices come in all flavors—from simple numeric pagers to sophisticated wireless wonders. Check out Chapter 10 for details.

Modem line tester

If you use a modem on the road, don't just plug it into any old wall jack. If you run into digital lines, your modem will encounter higher electrical current over the phone lines than it's designed for. The result—instant death for your modem. Coming to the rescue are devices that test the telephone port before you attach your modem. IBM makes one (800-388-7080). At less than $30, it sure beats the cost of replacing a dead modem on the road. Some of the newer PC modems come with a built-in line tester.

Digital/analog converter

Since analog telephone devices are not compatible with digital lines, you must use modems and other telephone equipment specifically designed for digital lines, or get a converter from Unlimited Systems, Inc. (800-275-6354) or TeleAdapt, Inc. (408-370-5105).

Tone dialer

If you encounter a dial phone (or a phone that looks Touch-Tone but doesn't deliver true tones), you'll want a tone dialer. This converts pulse tones into the tones you'll need for accessing automated phone features like voice mail and bank by phone.

Spare parts kit

Keep a small bag or shaving kit pre-packed with spare batteries (wrapped in foam or a cloth bag), a phone line extension cord, three-prong adapter, duo RJ-11 connector, screwdriver, small flashlight, extra diskettes, and the like.

Recognize This Problem?

Phone plug adapters

If you travel outside the United States, you'll need a variety of devices to adapt to the over 40 different types of phone plugs you might encounter. TeleAdapt sells a kit that contains a set of customized foreign telephone plugs, as well as some other handy converters. Magellan's catalog lists the common plug type needed for 150 countries (800-962-4943).

Handset coupler

Can't find an RJ-11 plug? You won't have to worry if you carry a handset coupler (aka acoustic coupler). It looks like a phone receiver with the ear and mouthpieces replaced by rubber cups. You place your telephone handset into the cups, connect a telephone wire between the coupler and your modem, and dial out.

Using a coupler lets you avoid having to scramble under a client's desk to plug in. And a coupler is especially useful if you travel to a variety of countries, or if you frequently need to connect using a payphone. You can get one capable of transmitting data at up to 24,000 bps, plenty fast.

Power for the road
Batteries are the life force of the itinerant worker. Buy an extra battery or two, keep it charged and carry it with you. If you don't, you'll kick yourself when you run out of power during that all-important call or when your modem blips out halfway through a crucial download.

Charge up batteries whenever you can. Make a habit of battery charging so you won't forget. For instance, get in the routine of doing it immediately upon arrival at your hotel room.

Battery chargers range from small portable trickle chargers to desktop-sized rapid chargers. If your equipment uses nickel-cadmium batteries, get a charger that has a discharge button so you can fully cycle the battery. Otherwise, your battery life will diminish every time you charge. This is caused by a strange but true phenomenon called battery memory build-up. The cell capacity decreases when batteries are only partially discharged and then recharged. For longer life and less memory loss, get either nickel metal hydride (NIMH) or lithium-ion (LISB) batteries.

Warning: Be careful when packing or toting spare batteries. The acids and heavy metals inside make them unstable. They're prone to electrical mishaps such as shorting out. If they're jolted too much, and if they're sharing a bag with loose metal parts (such as spare change or mini-screwdrivers), an accidental spark could start a fire. Avoid meltdown. Pack safely.

Solar power
How about a solar battery panel installed in your briefcase so you can charge up while taking an alfresco lunch or waiting for a flight? These featherweight solar panels can recharge

either from direct sunlight or via indirect artificial lighting. A special plug, similar to a car's cigarette lighter, transfers the stored energy to cellphones, laptops and other gear. Some cases come with solar panels preinstalled. If you want to convert your current case to solar, contact Keep it Simple Systems, which manufactures solar panels for PowerBooks and Newtons (800-327-6882).

Tips for working on the road

Before you leave
✔ Check your calling card. Your long-distance carrier may be levying a surcharge for each calling-card call you make. These range from 50¢ to 90¢ a call. Shop around for a long-distance company that charges no fee. Often, your best deal will come from a long-distance reseller. Many of them offer no-surcharge calling cards.

✔ If you usually make lots of long-distance calls, book reservations at a hotel that doesn't charge call access fees. Many hotels in the Sheraton, Stouffer and Hilton chains are of this type.

✔ If you make a lot of calls back to the home office, get a personal 800 number. Sometimes you can get an 800 number free if you purchase others services. Usage fees range around 20¢ to 25¢ per minute.

✔ Get a fax mailbox if you need access to your faxes on the road. Many phone companies, pager services, and mobile phone services offer this capability.

✔ Frequent travelers might be better served by a follow-me-anywhere number. Your calls can be forwarded to any number that you can directly dial, including cellular phones, fax machines, pagers and voice mail. You can sign up for this service with many local telephone companies as well as with national services such as Wildfire (800-WILDFIRE), Avox (415-623-3016) or MCI One (800-4MIC ONE).

✔ For international travel, arrange for an international access calling card such as AT&T's USA Direct, MCI's Call

USA and Sprint's Express service. Then, when you're abroad, you call your carrier's local access number. This connects you to a U.S. operator, who places the call for you. If you don't speak the language of the country you're in, this service is invaluable. Charges appear on your office phone bill, not your hotel bill. Good only for calls back to the United States. You could save up to 40%.

In the air

✔ On your next flight, you might be tempted to use an air-to-ground phone link such as AirOne, a communications system that has a standard phone jack in the handset of its in-air phone. You can plug in a portable modem to access e-mail, send faxes or dial up an online service once you're in the air. Watch those charges however. AirOne charges run about $2 a minute.

✔ Expecting an all-important phone call but can't miss your plane? Some airlines, such as Delta and USAir, offer the ability to *receive* incoming phone calls. Look for Airphone, a service that allows you to activate an on-board telephone located on the seat back in front of you. Calls are pricey: the minimum charge for a domestic call is around $7.

✔ Seasoned air travelers prefer seats on the window, facing the bulkhead, or seats on an exit row the best. Reason: you don't have to move your portable office setup every time a seatmate decides to get up.

✔ Don't boot up your computer, use a cellular phone, or turn on your portable fax machine until the plane reaches at least 10,000 feet. This is to ensure that the electronic emanations from your equipment don't interfere with cockpit controls.

At your hotel

✔ Look for hotels that serve business clients. They'll be equipped with the connections and furniture you'll need. For example, Marriott now offers rooms with two power outlets, a fully adjustable ergonomic chair, a PC modem jack, a mobile writing desk and task light. Others provide fax machines, two-line telephones and in-room data ports.

✔ Use a credit card when making long-distance calls from your hotel. If you don't, you'll end up paying about 40% more for each long-distance call. Hotels often charge a fee for using your phone credit-card instead of their phone service. These fees, called access fees, range anywhere from 50¢ to $1.25 a call.

✔ Save money on hotel long-distance by hitting the pound sign between each long distance call instead of hanging up. This will give you a new dial tone and save you from having to reenter your credit card number. More importantly, you'll avoid paying separate calling card access fees.

✔ When using a modem, ask the switchboard to put your call through as a direct call. Otherwise, you may not be able to get a true dial tone and your fax and modem calls will never reach their destination.

✔ Find out how to get direct dial access to your long distance carrier and then use the access code before every call or group of calls.

✔ A telephone debit card, or prepaid phone card, might help you save on hotel long-distance charges. These prepaid cards allow you to make nationwide calls at a flat rate. You can buy them in denominations of $10, $25, etc. To use a debit card, you call the 800 number listed on the card, enter a PIN code and dial the desired number.

Technologies for telecommuting

Software and hardware developers continue to come up with new products that can be used for workgroups. Some of the most useful:

Desktop document conferencing

This technology allows two people, separated by a phone line, to view and discuss the same file on their computer screens. Based on using a whiteboard in a meeting room, the two parties can draw on the screen using a mouse pointer, add typed comments, and paste in new numbers, text or images—all while talking about the work over standard phone lines. You'll need two copies of the document sharing software (one for each end of the conversation) and two phone lines (one for voice and one for data). In addition, each computer must have a modem—the faster, the better.

Crosswise Corporation's Face to Face (408-459-9060) is perhaps the most versatile conferencing software. Face to Face works with any software. It can communicate Mac to Mac, Windows to Windows or across platforms. If the person on the other end of your connection doesn't have Face to Face, you can give (yes give!) her a copy of Listener software, which allows her to participate in a conference. Listener software can also be downloaded from CompuServe or the Internet. See Resources for other companies.

Web conferencing

If you and your teammates have Internet access, you can share ideas and work on documents together via the Internet. Several firms produce Web-collaboration software that will let you cut-and-paste information between computers, share applications, and transfer files. You start a Web conference by typing in the person's IP address or clicking on a name in an Internet phone server. I've included a list of vendors in the Resources section at the end of this chapter.

Remote control

Using remote access software, you can dial up your company's mainframe or your personal computer at work and operate it as though it was sitting right in front of you. You'll want to use the fastest modems you can for this—one at each end. If you use a slow modem, you'll be frustrated. At 2400 baud, a full text screen takes about 5 seconds to display and a screen full of graphics takes about a minute. At 28.8, you can update a text screen in about quarter of a second and a graphics screen in about 5 seconds.

Using remote control software, you can update files, check your e-mail, print reports, synchronize files, and use office-based applications just as though you were in the office. One such package is Close-Up from Norton-Lambert (805-964-6767). Here's how it works:

On the days when you telecommute, start up Close-Up, call in to the host computer at the office and promptly hang up. Close-Up immediately calls your computer back. This is a security feature that prevents others from accessing the system. You can then log onto your company's network just as though you were down the hall. It will take just a few minutes to pick up your e-mail, leave messages for coworkers, and download needed files. At the end of the day, you can call again and synchronize both computers with completed work. See *Resources*, at the end of this chapter, for other remote control software vendors.

Remote power switch control
In order to access your computer at work, you either have to leave it on (and hope that some economy-minded soul doesn't turn it off) or get a power control device such as Remote Power On/Off (Server Technology, 800-835-1515). This device sits between the office modem and the electrical outlet. You program the number of rings (1, 6 or 12) that will wake up your computer. Then, to turn it on, you just call and wait until you've reached your ring threshold. Your computer stays on during your remote session and then turns itself off after the call is hung up. A remote that works with most Macs is PowerKey Remote (Sophisticated Circuits, 206-485-7979).

Node on the road
If you want to access your entire office network from afar, and act as if you're just another computer in the office, ask your network administrator the help you become a road node. Companies that make remote node software include Shiva's NetModem and LANRover products (800-458-3550) and Apple's ARA (800-510-2834) which is built-in to all Apple PowerBooks. Using one of these with an Ethernet-based modem, you can sign on to your office network and operate as though you were down the hall, rather than several states away.

Desktop video

Computer-based videoconferencing is a relatively new technology. It's perfect for situations that require demonstrations or face-to-face meetings. Desktop video is being used by companies to cut down on the amount of travel required. It's being put to use in a variety of interesting ways. Some examples:

- Students at Georgia Tech attend a Virtual Job Fair, including video interviews with corporate recruiters.

- Doctors at Kaiser Permanente hold noon-hour video meetings with their counterparts in other clinics. The company reports saving $1.5 million a year in physician time.

- Visitors at New York's famed Waldorf-Astoria Hotel can now book a room with built-in videoconferencing capabilities.

- Buyers at Minnesota-based Dayton Hudson Corporation demonstrate the latest fashions and merchandise over a video link with store managers all over the U.S.

- The Mortgage Network in Boca Raton, Florida provides video links between prospective home buyers and loan officers. They complete paperwork online and can have a loan approved—all in about ten minutes.

- Execs at Converse Inc. headquarters in North Reading, Massachusetts, use video meetings to iron out design and production problems. The result: the shoemaker is getting products to market faster.

- Circuit judges in California's Central Valley now hold video arraignments, saving cash-strapped county governments the cost of transporting prisoners.

- Produce buyers in Japan select California produce via videoconferencing. The images of fresh-picked strawberries, artichokes and other fruits and vegetables are beamed over ISDN lines.

- Hewlett-Packard reports that videoconferencing speeds product development by 30%. Another happy result: meetings are shorter.

How it works

The set-up consists of a small camera mounted on top of a monitor, one or more video adapter cards for the computer, a modem or ISDN adapter, software to run the program, a microphone, and a speaker or telephone handset. The software often comes with shared whiteboard or document collaboration software. When a videoconference is taking place, two computers are connected either by a standard analog telephone line or an ISDN line.

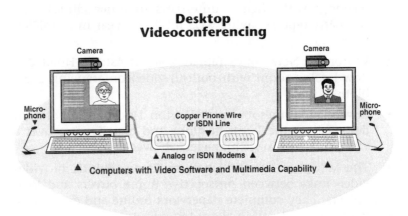

Desktop Videoconferencing

Camera

Camera

Micro-phone

Micro-phone

Copper Phone Wire or ISDN Line

▲ Analog or ISDN Modems ▲

▲ Computers with Video Software and Multimedia Capability ▲

Currently, most systems support point-to-point teleconferences, limiting your conference to two computers. To set up a conference, you launch the video software and dial the other computer by clicking on a photo in your graphical phone book. Then start any accompanying application you may want to view (such as a spreadsheet), seat yourself before the tiny video camera, and you're on stage. Your image (usually just head and shoulders) appears in a small box on the other screen.

Analog desktop video is not quite ready for prime time. The video quality is poor, with fuzzy images, jerky action and delayed audio. That's because analog connections support about 5 frames a second (versus 30 frames per second on your TV set). ISDN connections can double or even triple the number of frames per second, giving you acceptable video quality. The better video systems combine three ISDN lines for full-motion video quality. See *Resources*, at the end of this chapter, for a list of desktop video vendors.

Beware: Some desktop video systems are proprietary, which means that both conferees must use the same vendor's equipment. Make sure that the video products you buy adhere to the International Telecommunications Union standards—H.320 for video boards; H.324 for software.

Want to videoconference with your team or attend a long-distance training class? Check out group videoconferencing services. Many telephone companies and service bureaus provide group video services, which cost around $40 per hour per participant (plus long-distance charges) and can accommodate several locations. Each person dials into a toll-free number to join the videoconference. Participants can even dial in from a standard telephone for a voice-only connection. Works worldwide—anywhere there's an ISDN phone line.

If you don't have desktop video capability, but need to conduct an occasional videoconference, consider renting. Some office support companies and teleworking centers offer video-conferencing facilities. For example, Kinko's copy centers (800-743-2679) provide room-based videoconferencing at many of their branches.

Resources

Books
The Telecommuter's Advisor
by June Langhoff
Aegis, Newport, RI, 1996
An easy-to-follow roadmap to working from any remote location. Subjects include: designing your home office, selecting equipment, coping with e-mail, understanding modems, handling fax, managing messages, groupware, remote computing, cellular, paging, international communications, desktop video, working securely, and troubleshooting.

Making Telecommuting Happen: A Guide for Telemanagers and Telecommuters
by Jack M. Nilles
VNR Computer Library, 1994
This is the authoritative book on telecommuting from the organization's point-of-view. Nilles, who coined the term telecommuting over 20 years ago while stuck in a Los Angeles traffic jam, consults regularly with corporations and government agencies on telecommuting policy.

The Virtual Office Survival Handbook: What Telecommuters and Entrepreneurs Need to Succeed in Today's Non-Traditional Workplace
by Alice Bredin
John Wiley & Sons, New York, 1996
A comprehensive guide for surviving and thriving in a home office, mobile office, or any other non-traditional workplace.

The Telecommuter's Handbook: How to Work for a Salary Without Ever Leaving the House, 2nd Edition
by Brad and Debra Schepp
McGraw-Hill, New York, 1995
Packed with useful information for setting up a telecommuting relationship with your organization. Lists 50 jobs best suited for telecommuting and 100 companies with telecommuting programs.

Working From Home: Everything You Need to Know about Living and Working under the Same Roof, 4th Edition
by Paul and Sarah Edwards
Tarcher/Putnam, New York, 1994
This comprehensive guidebook is loaded with tips on setting up a home office and organizing for success.

Newsletters
Telecommuting Review
Gil Gordon Associates
10 Donner Court
Monmouth Junction, NJ 08852
908-329-2266
A monthly newsletter directed primarily at employers. Includes case studies, technical advice, legal and regulatory

developments, transportation issues, employee relations, and supervisory topics.

Associations
TAC: The International Telework Association
204 E. Street N.E.
Washington, DC 20002
202-547-6157
Internet: *www.telecommute.org*
A excellent group for networking and getting information on other companies and their telecommuting experiences. Membership includes subscription to a newsletter, quarterly teleconferences, and access to a comprehensive bibliographic service. Chapters exist in several major metropolitan areas.

Online support
CompuServe
The Working from Home Forum has a very active Telecommuting section. I'm the section leader in the telecommuting forum and look forward to welcoming you. To get there, type GO WORK.

Prodigy
Prodigy has a Telecommuting Interest Group Page on its webserver. You'll find it by using the command words JUMP TO TELECOMMUTING.

Remote control software

Carbon Copy	Close-Up
Microcom	Norton-Lambert
800-822-8224	805-964-6767
LapLink	ReachOut
Traveling Software	Stac Electronics
800-343-8080	800-279-7822
Timbuktu	pcAnywhere
Farrallon Computing	Symantec
800-344-7489	800-441-7234

Document conferencing software

ProShare
Intel
800-747-9060

ShareVision
Creative Labs
800-998-5227

Face to Face
Crosswise Corp.
800-747-9060

Omnishare
Hewlett-Packard
800-752-0900

Web conferencing

Look@Me
Farallon
510-814-5000

NetMeeting
Microsoft
800-426-9400

CoolTalk
Netscape
415-937-2555

Internet Conference
Professional
VocalTec
800-843-2289

Desktop video vendors

ProShare
Intel Corp.
800-538-3373

VIVO
VIVO Software
800-848-6411

ShareVision
Creative Labs
800-998-5227

PictureTel Live
PictureTel
800-716-6000

Vistium Personal Video
AT&T
800-VIDEO-GO

QuickTime Conferencing
Apple Compute
800-776-2333

CU-See Me
White Pine Software
603-886-9050

VideoPhone
Connectix
800 950-5880

15...Your Phone Bill
●●

J. Paul Getty, the fabled oil zillionaire, solved his phone bill problems by installing pay phones in almost every room of his mansion. You may not want to go to such extremes but, with a little effort, you can make changes that could improve your bottom line, too.

Getting a handle on your phone costs requires attention to detail. The time you spend in analyzing your telephone bills and planning for improved telephone efficiency will pay off in increased savings in time and money. This chapter discusses how to save money on long-distance charges, how to compare long-distance calling plans, and gives you loads of tips on how to control phone costs. These suggestions will get you started. The rest is up to you.

Save money on long-distance
Do you know much you pay for an average minute of long-distance calling? A recent survey showed that 88 % of the respondents knew what they paid for a gallon of gasoline and 86% knew the price of a loaf of bread, but only 38% knew the per-minute rate of a long-distance call. Time to change all that.

Types of calling plans

Just about every business around can save money by participating in a calling plan instead of opting for the basic long-
distance rate. There are a wide variety of calling plans, but
they usually boil down to one of five types:

Volume discount

These plans calculate your discount, ranging from 10% to
50%, based on your total monthly long-distance charges.
You usually have to meet some sort of minimum amount
(often $25 or more) before the discount kicks in. Calls are
charged on a per-minute and distance-basis.

Consider a volume discount if you make calls at varying
times of the day, evening and weekend, and make enough
calls to meet the discount threshold. Be sure to compare
these discounts against other plans, however.

Time-of-day discount

Your charges are less if you call during off-peak times.
Some plans have three calling periods: business hours,
after business hours (often from 5 p.m. to 8 p.m.) and
nights/weekends. Be sure you understand the carrier's def-
inition of business hours. Some programs define it from 7
a.m. to 7 p.m.

These discounts are best for telecommuters and solo busi-
nesspersons who can manage to make the bulk of their
long-distance calls after the normal business day.

Who you call or where you call discounts

Many of these discounts are based on a calling circle, or a
specific state or country that you call most often.

If you have a very predictable telephone pattern and most
of your calls are to specific phone numbers or area codes,
you may save with this type of plan. Otherwise, leave it
alone.

International plan

Most of these plans identify a country of your choice, give you some kind of volume discount, and have a monthly fee.

These plans are best for organizations that spend the bulk of their long-distance budget on calls outside the United States.

Flat-rate plan

These discounts are the easiest to calculate and budget. Some plans charge flat rates only during specific times of day.

Consider a flat-rate plan if a high percentage of your long-distance calls are coast-to-coast or exceed 1,000 miles. Most flat-rate plans charge a monthly fee. Be sure to factor that in to your calculations.

Choosing a plan

Even if you have local business service, you don't have to sign up for a business long-distance plan. You should evaluate both residential and business plans when shopping for a service. The Telecommunications Research and Action Center advises you to pick a residential calling plan if you usually spend $75 or less for interstate long-distance each month. If you spend between $75 and $200, you should consider either a residential or business calling plan. If your call volume exceeds $200 a month, you'll want a business calling plan.

If possible, get a calling plan that provides fractional billing. This measures calls in six-second increments instead of full-minute blocks. If you make a lot of short calls (calls where you leave voice messages or fax calls that are unsuccessful on the first try), you could save up to 35% using fractional billing.

Analyze your calling behavior

To pick a cost-effective plan, you'll need to analyze your long-distance phone bills. Here's a strategy that works:

1. Assemble a few month's worth of long-distance bills. Chart the numbers on a matrix or spreadsheet.

2. Determine your per-minute rate. Add up the total minutes on your monthly bill and divide this into the dollar amount on your bill to find your average cost per minute.

3. Look for patterns. Does one area code get the largest percentage of your calls? What time of day are most of your calls made? Do you call the same few numbers?

4. Based on your calling pattern, narrow your choices to one or two plan types.

5. Call long-distance carriers and resellers to do some comparison shopping. Provide your average monthly usage, the total number of calls, and the total number of minutes. Ask each company to quote the best service for your needs. Ask them to provide their per-minute rates.

6. Set up a spreadsheet or matrix to compare the plans. Plug in the numbers, note any exceptions, deals and fees.

7. "Test drive" the service that looks the best on paper. You do this by using the company's five-digit access code before you dial the desired number. For example, if you wanted to try Sprint service, simply enter 10333 before you dial the desired number. You'll be billed for the call on your local phone bill.

8. Read the fine print on the service contract. Pick your plan. Avoid making long-term commitments, unless the discount is fabulous.

9. Rates, services and features are constantly changing so you should reevaluate your plan every few months.

Shopping for a plan

Shop around for discount calling plans. Don't restrict yourself to just the big carriers like MCI, Sprint & AT&T. You may get a top rate from a reseller (also called a *switchless reseller, rebiller* or *aggregator*). There are over 300 long-distance resellers in the United States. They lease phone lines from the major carriers and resell them at a discount to small businesses like you. It's sort of like a co-op bulk buying service.

Many resellers will allow home-offices to combine billing for residential and business lines to qualify for discount plans.

Choosing a Reseller

Be sure to ask the reseller lots of questions. Here's a convenient list:

❑   Which carrier's service are you reselling?

❑   How long have you been in business?

❑   Do you provide 24-hour customer service?

❑   Are you certified with the state Public Utility Commission?

❑   Are you a member of the Telecommunications Resellers Association?

You can find a reseller by contacting the Telecommunications Resellers Association, (202-835-9898).

The Telecommunications Research and Action Center (TRAC) publishes a newsletter called *Tele-Tips*. This newsletter contains billing plans, peak and off-peak rates and other features offered by a number of long-distance companies. You can get a copy of the newsletter by sending a self-addressed, stamped envelope and a check ($5 for residential; $7 for business) to:

TRAC
P.O. Box 27279
Washington, DC 20005

Comparing plans

It's tough to compare long-distance plans because the features and pricing structures are so variable. It's like comparing apples to oranges. But just as fruit can be compared on the basis of caloric value or the amount of vitamin C it contains, calling plans can be similarly evaluated.

One of the best comparisons is the average cost-per- minute. You might want to break this down to the cost-per-minute by time of day or by distance. It's also useful to know what fees and minimums are required, whether the calls are billed on a six-second or per-minute basis, and what hours constitute normal business hours.

Example: June picks a calling plan

In an effort to practice what I preach, I recently analyzed my own long-distance behavior. I used three months worth of long-distance bills and totaled the minutes.

	July	Aug	Sept
Total bill	$35.12	$41.13	$66.83
Total # minutes	186	218	355

Then I broke the bill down into component parts: how many minutes were spent during business hours, night hours (5-11 p.m. Sun-Thurs) and weekend hours. I also looked at how many of the minutes were spent on interstate versus intrastate calls.

# Mins. daytime	122	153	217
# Mins. night	34	37	84
# Mins. weekend	30	28	54
# Mins. interstate	139	170	299
# Mins. intrastate	47	48	56

Next, I calculated my average monthly bill and the average cost-per-minute. I also totaled each column. Now I can see some patterns emerging.

	July	Aug	Sept	Total
Total bill	35.12	41.13	66.83	$143.08
Total # minutes	186	218	355	759
Avg. cost/minute				18.9¢
Avg. monthly bill				$47.69
# Mins. daytime	122	153	217	492
# Mins. night	34	37	84	155
# Mins. weekend	30	28	54	112

	July	Aug	Sept	Total
# Mins. interstate	139	170	299	608
# Mins. intrastate	47	48	56	151

65% of my calls take place during business hours. Of the remainder, 20% take place between the hours of 5 p.m. and 11 p.m., Monday through Thursday, and a mere 15% take place on weekends or between the hours of 11 p.m. and 8 p.m. Eighty percent of my calls are out-of-state calls. These are usually billed at a higher rate than in-state calls.

So, armed with my new self-knowledge, I made some calls and assembled information on three types of calling plans.

The first type I checked were plans based on a volume discount. Here are three typical plans:

Volume discount plans

	Plan A	Plan B	Plan C
Minimum monthly bill?	$10	$10	none
Monthly fee?	none	none	$1.95
Volume discount?	yes	yes	yes
Discount up to $10	none	none	15%
Discount $10-$24.99	10%	25%	15%
Discount $25-$74.99	20%	25%	15%
Discount more than $75	30%	25%	15%

I can calculate my projected savings by multiplying my average monthly bill by the discount amount. If there's a monthly fee, I'll have to take that into consideration, too. Sample calculation (Plan A):

My average monthly bill	$47.69
- discount (20%)	9.54
+ fee (if any)	
Projected monthly bill	$38.15

Next, I gathered information on three plans that based charges on time-of-day.

Time-of-day plans

	Plan D	Plan E	Plan F
Minimum monthly bill?	none	none	none
Monthly fee?	none	none	none
Daytime cost/min.	19¢	24¢	25¢
Nighttime cost/min.	14¢	10.5¢	10¢
Weekend/cost/min.	12¢	10.5¢	10¢
Daytime definition	7-7 MF	8-5 MF	8-5 MF

The calculations are a bit more complex for this plan. I multiply the average minutes spent in each category by the cost-per-minute for that category.

Sample calculation (Plan D):

Average daytime minutes	164.00
x cost/minute	.19
Projected daytime charges	$31.16

I do this for each category and sum the projected charges to determine my projected monthly bill.

Sample calculation (Plan D):

Projected daytime charges	$31.16
Projected nighttime charges	7.28
Projected weekend charges	4.44
Projected monthly bill	$42.88

I decide to check two more plans before doing the grand comparison, so I go for flat-rate plans with a monthly fee. I liked the idea that it would be easier to project my costs. And they looked like I might save even more money.

Flat-rate plans with monthly fees

	Plan G	Plan H
Minimum monthly bill?	$25	none
Monthly fee	$3	$5
Daytime cost/minute	22¢	20¢
All other times/cost/min.	10¢	10.5¢

I multiply the average minutes spent in each category by the cost-per-minute for that category and add the monthly fee to determine the projected monthly bill.

Sample calculation (Plan G):

Projected daytime charges (164 x .22) 36.08
Projected nighttime charges (89 x .10) 8.90
+ Monthly fee   3.00
Projected monthly bill $47.98

Which plan works best for me?

Plan	Cost	Savings
Plan A - 20% discount	$38.15	$9.54
Plan B - 25% discount	$25.77	$11.92
Plan C - 15% discount	$42.49	$5.20
Plan D - 19¢ daytime rate	$42.88	$4.81
Plan E - 24¢ daytime rate	$46.68	costs 99¢ more
Plan F - 25¢ daytime rate	$49.90	costs $2.21 more
Plan G - 22¢ daytime + $3	$47.98	costs 29¢ more
Plan H - 20¢ daytime + $5	$47.15	54¢

The final result: My original bill was $47.69. Plan B, the one with the 25% discount for charges between $10 and $75, gives me the best savings. For now, anyway.

Tips for controlling phone costs

I've assembled a bunch of tips that can do wonders in helping you get a handle on your phone bills. Try some out. You'll be glad you did.

✔ Curtail 411 calling

Be careful about calling directory assistance. It's not free any more, although some plans allow two or three free calls a month per line. Depending on your area, a call to directory assistance costs from 50¢ to 85¢. The average is 65¢. At that rate, just one call a day racks up around $20 a month.

If you frequently use directories, get yourself a disk-based or CD-ROM directory. These useful computerized directories contain business names, addresses and even fax numbers. You can look up information by name, phone number, or even business type.

There are several good directories available. Check out *Select Phone* (ProCD 800-992-3766), or *16 Million Businesses Phone Book* (American Business Information, 800-555-5999).

If you have access to the Internet, check out the Nynex Interactive Yellow Pages *(http://www.bigyellow.com)*, which lets you search yellow pages listings anywhere in the U.S. by state, name, or business type. The World Yellow Pages Network *(http://wyp.net)* has over 115 million residential and business listings in the U.S. and Canada.

✔ Call off-peak

If possible, save up your long-distance calls for off-peak hours. I try get to work early to make long-distance calls to companies back east.

Warning: Most long-distance companies will bill your call at the day rate if any portion of the call extends beyond the off-hours. For example, if you place a long-distance call at 7:45 a.m., and hang up at 8:01 a.m., the entire call will be billed at the daytime rate.

✔ Fax at night

Program your fax modem or fax machine to send your faxes at night when phone rates are lowest. You can save 20% to 50%, depending on the distance involved and your calling plan. To save even more, use e-mail instead.

✔ Shorten your calls

Limit the time you spend on the phone by offering to send (or fax) a proposal or price list. Then follow up later with a quick phone call. While you may spend 10 minutes on a voice call, you may be able to transact the same business with a one-minute fax. If you pay 25¢ a minute for a daytime long-distance call, you'd save $2.25. It all adds up.

Keep a clock by your phone or get one of those phones with a built-in timer. You'd be surprised at how much you can save by watching the clock.

✔ Code your calls

Writers, lawyers and many other professionals have to track reimbursable time spent on the phone. To do this:

1. Keep a diary of all telephone calls. However, this is tedious and can be easily forgotten.

2. Invest in a call-reporting device such as CallCost from Hello Direct.

3. Use call accounting codes provided by your long-distance company.

However you do it, call accounting is useful for keeping track of your time on the phone. If you use coding supplied by your long-distance company, you could sort your calls (and faxes) by client name, or identify types of calls (sales, billing, publicity). Then, when your bill arrives, you have these tidy codes next to each call. Many long-distance carriers and resellers provide this service for free or for a small monthly fee.

Another advantage of call codes is that they tend to intimidate your staff into more appropriate calling behavior. Some companies report that using accounting codes virtually eliminated personal calls and reduced phone costs by 10% to 40%.

For more information, call:

- AT&T Small Business Services (800-222-0400)
- MCI (800-727-5555)
- Sprint (800-877-7746)

✔ Control phone abuse

To stop unauthorized calling by employees, get an inexpensive call-blocking device that you clip onto the phone line cord. You can pick one up for less than $12 at your local electronics store. Or you can get a blocker with some intelligence that will allow you to restrict calls selected by area code, prefix or even by specific number.

An alternative is to install a personal pay phone in your employee lounge and require that all personal calls be made on that phone. The revenues from these calls could go into a group coffee or snack fund, to blunt any negative effect on employee morale. Hello Direct carries these phones.

✔ Consider residential service

If you don't need a business directory listing, you may be able to save a bundle by using residential, rather than business service. Several telephone companies—including GTE, US West, Sprint and BellSouth—are allowing residential phone lines to be used for business. More and more state Public Utility Commissions are allowing this change in tariff regulations. Examples of savings:

- Minnesota: US West charges $47.63 for a business line and only $18.20 for a residential line. In a year, you'd save $353 a line.

- California: Pacific Bell charges $14.93 for a business line and $11.25 for a flat rate residential line. In a year, you'd save $44 a line plus all per-minute usage charges in your calling area.

✔ Share a line

If your call volume is low, and service to your customers won't suffer if your callers experience an occasional busy signal, look into line-sharing. You can purchase a switch ($30 on up) that will let you share a line between phone, fax and answering machine. Some equipment comes with built-in switching capability or works with Distinctive Ring services, provided by your local telco. You can find line switches in telephone and electronics stores.

✔ Use a prepaid calling card (sometimes)

Buy flat-rate long-distance using a prepaid calling card (also called a debit card). Cards come in denominations like $5, $10, $25 and the like.

These cards are easy to use. You dial an 800 number listed on the back of the card, enter a PIN number (also printed on the back of the card), and then the number you're calling. Your calls are charged against the card until the value is depleted. When the card value is nearly gone, an automated voice warns you that you have only 30 (or 60) seconds left to wrap up your call.

Prepaid cards have a high per-minute charge, often in the neighborhood of 45¢ a minute. So you don't want to use these cards for just any call.

Use a prepaid card when you would normally use your telephone credit card, but only when the access fee makes your credit card call cost more than the prepaid call. Access fees range from 30¢ to 90¢ a call. The average is 80¢. If your call is short and your distance long, you'll save money using a prepaid card over using your credit card, because you initially dial a toll-free 800 number, and pay no access fee. For example, compare the cost of a two-minute call from San Francisco to New York during the business day:

Calling card (27¢/min)	Prepaid card (45¢/min)
$.80 access fee	no access fee
.54	$.90
$1.34	$.90

You can purchase prepaid cards at stationery shops, electronics stores, even at your local corner deli. They make nice gifts. But be sure to calculate the per-minute cost before buying one. Check for expiration dates. Some expire within six months; others have no ending date. Some you can "recharge" by inputting your credit card number and selecting how many minutes you wish to purchase.

✔ Audit your phone bills

Carefully examine your phone bill each month. Believe it or not, you may be paying for lines or services that should have been disconnected long ago. Auditing is especially important if you took over your company from a previous owner. If you find a discrepancy, notify your phone company and arrange for a refund.

Get a copy of *Local & Long Distance Telephone Company Billing* by Michael Brosnan & John Messina, Telecom Library Inc. (212-691-8215). This book will help you decipher your phone bill, and it offers loads of useful tips for reducing your local and long-distance bills.

If your phone bill is complex (or you have a lot of lines) and you need help deciphering it, enlist the aid of a telephone bill auditor. Auditors are professionals who will examine your bills, identify discrepancies, and negotiate with the phone company for you. They usually charge a percentage of any refunds they obtain. You can find an auditor in the Yellow Pages under *telecommunications consultants* or call the Society of Telecommunications Consultants in Boca Raton Florida (800-782-7670) for a list of members.

✔ Tackle toll fraud

American businesses lose at least $5 billion a year from telephone fraud. As reported in *Teleconnect* ("Toll Fraud Terror," January, 1996), The average cost of just one incident of toll fraud is $44,000—enough to sink many a small business. Telephone hackers are especially fond of preying on small businesses because they're usually unsuspecting and unprotected. Unlike credit card fraud, where your liability is limited, phone fraud losses are the customer's responsibility. What to do?

If you have a phone system, consider installing a fraud detection system. These devices monitor your system and alert you to signs of tampering, such as lots of calls made during non-business hours, a higher percentage of credit card calls, or too many busy signals. Some devices will automatically take your phones down at the first sign of fraudulent activity. Toll fraud prevention doesn't come cheap. Equipment to monitor your system ranges in price from around $800 to well over $15,000. Contact your local phone vendor for assistance.

If you have a cellular phone, examine your bills extra carefully. Phone cloners and other crooks have many ways of charging their calls to your phone bill. For details, see *Cellular fraud* in Chapter 7.

✔ Use a personal number when out of the office

If you tend to make a lot of calls back to the home office when you're on a trip, a personal toll-free number may be cost-effective. Often you can get an 800 or 888 number free if you purchase other services from your long-distance provider. Usage fees average around 20¢ per minute.

✔ Arrange for follow-me-service

If your office has to call you a lot, investigate a follow-me-anywhere number such as Wildfire (800-WILDFIRE), Avox (415-623-3016) or MCIOne (800-4MCI-ONE).

✔ Control hotel phone charges

Unfortunately, many hotels seem to view their phone service as a for-profit center, rather than as a service for guests. Therefore, never use the hotel phone service for long-distance calls without checking their policy. Use your credit card instead. If you don't, you may end up paying about 40% more for each long-distance call.

Many hotels also charge an access fee (ranging from 50¢ to $1.25 a call) if you use your phone credit-card instead of their phone service. When making reservations, ask if the hotel charges calling card access fees. If they do, try to stay somewhere else. The Sheraton, Stouffer and Hilton chains don't charge fees.

Tip: You can save money on hotel long-distance by hitting the pound sign between each long-distance call instead of hanging up. This allows you to avoid paying a calling card access fee for each separate phone call.

✔ Beware of pay phones for long distance calling

Independent pay phones have been found to charge up to ten times the rate that you'd pay for a for a long-distance call from home. As reported in *U.S. News & World Report* ("Watch That Pay Phone", 6/26/95), an 18-minute call from Ramsey, New Jersey to New York City cost $21.32 on an independent pay phone. AT&T would have charged $5.53.

To avoid these exorbitant charges, dial 00 before you call and ask that the long-distance operator place the call through your own long-distance provider.

✔ Dial toll-free as much as possible

This one is pretty obvious. If you need to find an 800 or 888 number fast, call 800-555-1212 or check out AT&T's directory on the Web (*www.tollfree.att.net/dir800*).

✔ Dial direct

When you dial direct, your call will cost up to 60% less than the cost of an operator-assisted call. When you are in your office, your phones are automatically programmed to use your preferred long-distance carrier. However, if you are out of the office, you must specify your carrier or use the one that is preprogrammed for that particular phone.

Call your long-distance company and ask for its direct-dial access code. Use the access code before every call or group of calls whenever you're dialing from a phone outside your office. Some common codes are:

    10222 = MCI
    10288 = AT&T
    10333 = Sprint

Cost-saving tips from a pro

Texas-based Denise Munro is a principal with the Cost Reduction Group. Her firm specializes in analyzing telephone bills for companies with telephone costs of $2,000 a month or more. In a recent audit, she negotiated annual costs savings of $100,000 for a client who had inherited a hodge-podge of services through mergers and acquisitions. Some of the services on their bill were unused; others had been previously disconnected were not removed from the bill. Such mistakes are not rare. According to Denise, "It is estimated that 70 percent of all business telephone bills are wrong, usually in the phone company's favor."

Though most of her customers have substantial telephone bills, her suggestions are useful for organizations of any size. Her advice:

❑ Invest in a good phone system. Phone systems today are nearly maintenance-free. Get one that lets you program it yourself. Ask your vendor to teach you how to set up and care for your system.

❑ Consider canceling your maintenance agreement. If you had two or fewer service calls last year, your maintenance contract wasn't paid for. The hourly rate for a service call is somewhere around $65.

   Note: If you don't have a service contract, your vendor will not give your priority service. If your communications are critical, you may need to keep the service contract just to get good response times.

❑ Discontinue any cable protection programs you may be paying for. These cost anywhere from $1.50 to $2.50 per line each month. If you are already paying for a phone system maintenance contract, you're paying twice. Remember, if your vendor installed the phone wiring, it is responsible for the work.

❑ If you experience phone trouble, call your vendor first, not the telephone company. Usually, its service trip charges are cheaper than the telco's. If the problem is with your lines, your vendor can coordinate repair.

❑ Be sure that your long-distance plan is billed in six-second increments. Call more than just the major carriers when comparing plans. The Big Three may not offer the best plans for small organizations. Look into service from other long-distance carriers (such as Allnet or LDDS Metromedia) or from a reseller.

❑ When selecting a long-distance carrier, ask for references and give those references a call. Get the plan in writing. Be sure that you are comfortable with service programs and include them in a written contract. Don't sign long-term agreements. You can always find a better plan.

❑ Know what you are paying for. Your telephone company can provide a detailed list of the lines, services, etc. for which it is billing you. Compare the list to what you know you have and question what you don't understand.

❑ If you find you are being overcharged, or do not know if you are, an auditing firm will pursue your case on a contingency-fee basis. It can also offer recommendations that produce additional monthly savings.

Resources

Long-distance carriers & resellers

There are over 300 long-distance companies in the United States and Canada. In addition, some areas of the country now allow local phone companies to offer long-distance service. To find long-distance services in your area, look in the Yellow Pages under *Telephone Companies* or *Telephone Communications Services*. Ask your business associates for their recommendations. You can find a reseller by contacting the Telecommunications Resellers Association, (202-835-9898).

16...Future Planning

●●●

For your organization to grow and prosper, you need a plan. Since this isn't a book on writing business plans, I'll leave that part of the planning up to you. However, your plan needs to include your communications needs. If you don't consider growth when you're setting up, you could end up with communications that hamper your development.

It's so easy to overlook your communications system. After all, phones tend to look like just part of the furniture. The only time you really think about your fax machine may be when it's out of paper. And when was the last time you seriously took a communications inventory?

This chapter is about planning for the future—for your business survival. This will include considering whether or not to add a LAN or EDI to your setup, some advice on disaster planning, and help in determining how many lines you need. I've included several examples of actual businesses and how they currently handle their telecommunications. And finally, I'll try to sum it all up for you.

Is it time for a network?

A Local Area Network (LAN) is a collection of computers and other equipment that is linked together using cable or telephone wire. There are even some LANs that communicate using laser, infrared, cellular, radio, or satellite signals. Most

LANs cover a small geographic area—often just a couple hundred feet.

Many small businesses could benefit from a LAN, but mistakenly think that LANs are only for larger organizations. Here are some of the things you could do if you had a LAN:

- Save money by sharing a printer or modem (fax, data or voice) among several computers.
- Maintain a company-wide calendar and organizer.
- Track projects online.
- Leverage your disk space; there's no need for everyone to keep separate copies of software, databases and files.
- Store all your company forms on the network and print only as-needed (frees up storage space).
- Work on files without having to duplicate them (or worry about which revision you're using).
- Connect incompatible computers like Macs, Windows PCs and DOS PCs.
- Easily access company information literature and forms from remote locations.
- Simplify and standardize backup.
- Save the time and duplication costs of printing interoffice memos. Replace them with e-memos.

Most LANs contain just a few parts: (1) Each computer has a network interface card (NIC); (2) each computer has some kind of software to connect the user to the network and to shift files and data from computer to computer; (3) everything is bound together by connecting cables, although there are a few LANs that operate wirelessly. These cables are either coaxial (like the stuff that runs to your cable TV), fiber-optic (expensive but powerful) or twisted-pair telephone wiring (just like the wires that carry your telephone signals). Because twisted-pair wiring is the least expensive, it's the most popular. No surprise there.

Types of LANs

Client-server

You dedicate one computer to be the *server*. This computer acts as a traffic manager and controls network activities. Usually, you'll need help setting up and maintaining a client-server network. Consider this type of LAN if you need high speed and use computer-intensive applications like CAD or desktop publishing. You can set up a client-server LAN for somewhere between $200 and $400 a workstation, not including the cost of the serving computer and the cost of having someone set it up for you.

Peer-to-peer

Each computer in the network is equal, therefore, there is no need to set aside a computer to serve as boss. Peer-to-peer networks, although slower than client-server networks, are relatively inexpensive and much easier to set up and maintain. You can even find peer-to-peer kits at your local computer store. You can set up a peer-to-peer network for about $200 to $400 a workstation. Look for Artisoft's Simply Lantastic and Microsoft's Windows for Workgroups.

Zero-slot

This kind of network requires no internal network interface cards. Instead, it connects through the serial or parallel port on each PC. Zero-slot networks are pretty slow. Consider using a zero-slot network if all you need is to link a couple of computers to a printer. Costs range between $50 and $90 a workstation.

Do you need a LAN?

If your needs are just file-sharing, you can probably get along just fine using *sneakernet* (walking down the hall with a floppy in your hand). If you want to share a printer or other peripheral, you might consider getting some kind of switch instead. However, if you want to automate your office, save on disk space, and simplify the paperwork, a LAN may be just the ticket.

Networking is destined to get simpler every year. If your office has Macintosh computers, you've had networking capabilities for years through the built-in LocalTalk circuitry.

Want more information? See *Resources*, at the end of the chapter.

EDI—another kind of network

Chances are your business uses a mix of high-technology communications such as fax and e-mail and no-tech approaches like paper invoices, NCR purchase orders and handwritten checks. You get an order, maybe via fax or over the phone, and type it onto a multipart form. One copy goes to the customer, one to shipping, one to accounting. Shipping types up another form and sends copies to . . . you know the drill. Invoices, transfer sheets, bills of lading, checks, inventory ledgers, manifests—some may be automated, most aren't.

Wouldn't it be great to have a system that lets you download orders over a network, which automatically generates a purchase order, invoice, and all the shipping documents? Well, you can. Just get connected with an EDI network.

Electronic data interchange (EDI) is an umbrella term for computer-to-computer business transactions of all kinds. It may involve a transfer of funds (e-money), generating a purchase order, creating an invoice, preparing a shipping request, or a number of other automatic functions involved with commerce. There are electronic networks for all kinds of industries including:

- Hospital supplies
- Gift shops
- Auto parts
- Insurance
- Bookstores
- Office supplies
- Electronic equipment

At least 40,000 American companies are involved in EDI networks today, and the number is growing rapidly. Some large organizations, such as Wal-Mart Stores, Ford Motor Company, and the Department of Defense, have made EDI mandatory for all their suppliers. The alternative is to be dropped from the approved vendor list. As organizations downsize and continue to cut costs, EDI will continue to grow. Here are some examples of how small-business people are joining the EDI world:

Aspen Press Ltd. of Chicago publishes children's books. In business just over three years, Aspen already bills over $1 million a year. How does it get high-volume orders in such a competitive field? By accepting orders and payments electronically. Using EDI, Aspen has been able to land large customers, such as JC Penney and Dillard Department Stores, which required their suppliers to tie into their electronic network.

Betsy Flags in Lake Bluff, Illinois produces—you guessed it—American flags. In order to sell to large retail chains such as K-Mart, Betsy had to convert to EDI. Now, once an order comes in, Betsy prints custom bar-coded labels to match the customer's specifications.

Mikana Manufacturing Company, a small tool-and-die shop in San Dimas, California, supplies parts for giants like Northrop Grumman and Bristol-Myers Squibb. Using high-tech tools like computerized ordering and job-tracking, Mikana can continue to meet its customer's needs. For Mikana, the choice was basically "automate or die."

Michael Finken, President of EDIS Inc., an EDI consulting firm in Greenville South Carolina, gives this advice to companies considering EDI:

❑ Acquire a copy of the *EDI Yellow Pages* from Phillips Business Information Inc. (800-777-5006). This lists organizations that are involved in EDI, broken down by industry type. You'll also find contact numbers and EDI consultants.

❑ Contact the Data Interchange Standards Association (DISA) (703-548-7005). Get on its mailing list. Attend one of its regional industry seminars to see what others in your industry are already doing.

❑ Decide what computer platform you are going to use. Make sure the computer is capable of handling your potential processing volume.

❑ Decide the level of your commitment to EDI. Are you just doing it to get on the vendor list of one particular customer or do you want to automate all your paperwork?

❑ Get a consultant's help to guide you through the maze.

Planning for disaster

Six out of ten businesses that suffer from a disaster severe enough to close them down temporarily never reopen. Don't be one of them.

The Geary Theater in San Francisco was home to the American Conservatory Theater (ACT) until 5:07 p.m., October 17 1989, when the proscenium arch crashed to the stage. A magnitude-7.1 earthquake had rocked the area and closed the theater. Fortunately, the stage was empty and no one was hurt. However, the company had to relocate all its productions for the entire season to other locations around town. One office suddenly became six.

According to Fred Reppond, ACT's telecom administrator, they were lucky. "We had just signed up for Centrex service. Although we were unable to use the theater, our phone system did not go down. Pacific Bell was able to connect the new locations so we could continue operations from six different sites as though we were all under one roof. We'd never have been able to do that if we'd had our own switch."

In case you're concerned, life is not over for the 80-year-old Geary. It was seismically retrofitted and reopened in early 1996.

Kerri McBride provides business and market planning services to clients from her home office in Boca Raton, Florida. Because she lives in an area that experiences severe thunderstorms, she is careful about protecting her equipment from electrical spikes, and has surge protectors on all computer equipment. In October 1994, lightning struck near her home. Her surge protectors were undamaged. However, she says: "I had a dead PC (motherboard, fax card, sound card, scanner card, and blown, unrecoverable hard drive) *and* a dead color deskjet, three dead phones, two dead TVs not to mention a cooked microwave."

What happened? As Kerri tells it, a big power surge "traveled into my home on the electrical line to the phone wires inside my home. It traveled over the phone wiring inside the walls, and entered my PC via the internal fax/modem port and made its way through just about every component I own."

Kerri was completely out of commission for five days. Fortunately, she routinely backed up files onto floppies so she didn't lose client files. But the time required to negotiate with insurance companies, unravel vendor warranties, purchase replacement equipment, reload programs, recreate customized templates, rebuild fax signature files, and the like, took its toll.

McBride offers this advice:

❑ Judge your surge protector by the value of the warranty guarantee. Don't buy one that doesn't cover the replacement of your system.

❑ Buy all your surge protectors from the same manufacturer. That way no one will argue about whose protector was the bad guy if a surge protector fails.

❑ Be sure you have computer hardware insurance (Kerri did) and know what the deductible is and what is covered.

❑ Check the policy for equipment still under warranty. Many companies won't replace items damaged by "Acts of God."

❑ Buy surge protectors with RJ-11 jacks for your phones and fax lines. And use them.

McBride's story is not a rare occurrence. It could happen to anyone. A vendor who specializes in used telco equipment crowed after a recent thunderstorm, "This last storm was great for business! I made $25,000 in lightning today!"

Even surge protectors may not protect you from a direct hit. Many experts advise that you physically unplug all equipment from the wall (including your phones) if you find yourself in the eye of a thunderstorm. Don't plug back in until the lightning strikes are at least a second (1,000 feet) away.

You may think that if you aren't located in earthquake country or thunderstorm alley, you've got nothing to worry about. Wrong! Disasters come in all sizes and varieties—from localized catastrophes like fires and broken water mains to regional disasters like floods and windstorms.

Disasters aren't always caused by Mother Nature. Sometimes humans are at fault, as with strikes, civil disturbances and sabotage. Then there are systems failures such as power spikes and outages, toxic spills, cut cables and leaky pipes. What would you do if a street construction crew accidentally cut through a cable that linked your phone lines to the outside world?

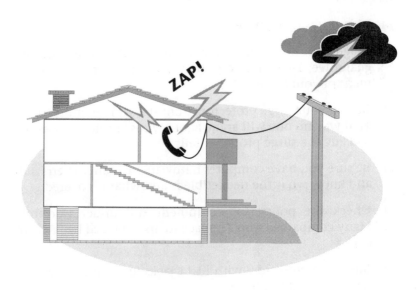

If there's a power outage, your phones won't ring and your lights won't light up if you are using an electronic phone system. Although you can make outgoing calls, incoming calls have no way of alerting you.

The folks at Dolphin Real Estate, a busy real estate office in Pacifica, California, wondered why the phones were so quiet one day. As co-owner John Doyle tells it, "We discovered that the power system for the phones was down. The only way agents could tell if a call was coming in was to pick up a phone and say hello. If someone spoke, we had a call. We had to endure this situation for three days before it could be repaired."

Since this situation is unacceptable, telephone system designers usually install either backup battery packs or a bypass system that allows at least some phones to ring if power is out. If you don't have this capability, get it.

Disaster-proofing your communications

✔ Set up a disaster communications plan for your organization. Determine how to forward your phones somewhere (answering service, voice mail, even employee's homes) so that your business can stay open. Create an alternate communications plan so your employees can find out when and how to get to work. Many voice mail services offer standby voice mailboxes to use in emergencies. Put the plan in writing. Hold disaster drills.

✔ Online computer services can provide an excellent back-up communications link in an emergency. In January 1994, when a major earthquake knocked down long-distance phone service in the Los Angeles area, Prodigy fielded more than 800,000 calls in a day.

✔ Mobile phones are a boon in emergencies. After Hurricane Andrew devastated South Florida in 1992, federal disaster teams set up emergency field offices using over 4,000 mobile phones to coordinate the recovery effort. Onsite insurance adjusters carried portable computers with wireless modems. They filed claims over a wireless system, and repair estimates and claims settlements were made rapidly.

✔ If you have 800 or 888 service, the number can be forwarded to a temporary location with just a phone call to your service provider. That way, orders and inquiries won't be lost.

✔ If you are unable to enter your business site for some time, check to see if you can arrange for a remote line (Remote Call Forwarding). The service will forward your calls to an alternate site, branch office, even your home.

✔ Encourage telecommuting. Though your employees may not be able to get to work because roads or transportation facilities are out, if they have an operable phone line or a portable computer, they can work from home.

✔ Don't rely on cordless phones alone. They need electrical power in order to operate. Tough luck in a brownout or grid failure.

✔ Make sure all your surge protectors have jacks for phone and modem lines.

Putting it all together

Figuring out the right communications mix that works for you can be a tough task. I thought that some examples of how other small businesses have set themselves up might help you sort it all out.

One line (needs another)

Dwight Kiyono, a freelance broadcast video engineer, operates out of his home office in El Cerrito, California. Kiyono has been getting along with a single phone line and an answering machine (Panasonic KX-T1423) for years. Because he performs services for a very specific client base, he advertises in an industry phone book rather than in the local *Yellow Pages*. This puts his number where it will get him work, saves him from having to field a lot of "Do you videotape weddings?" queries, and keeps his answering machine from filling up with calls from people who are "just shopping." It also allows him to subscribe to a residential line rather than a business line, at a lower cost.

Kiyono recently decided that he needs to add more communications capability. "My client base is growing and I need to handle a larger number of callers and be sure that no one feels neglected," he says. Within the next several months he plans to add a second phone line, paging service, and a fax machine. Dwight says he has lost jobs in the past because he wasn't able to return a prospective customer's call in time. "In my business," he explains, "when people call for production jobs, they're not just calling me. They're usually calling several people to assemble a crew for a shoot. The first person to call back gets the job."

Kiyono needs a fax machine for many reasons. His clients often want to fax maps showing how to get to a shooting location, or to transmit lists of special equipment requirements. The fax is also used to transmit electrical schematics and circuitry designs for multi-camera hookups, invoices, and reminders. Right now, Kiyono uses a fax service bureau a few blocks away from his house. He pays 50¢ a page for incoming fax; $1 a page for outgoing. He estimates that he spends about $10 a month at the fax bureau.

Kiyono estimates he'll spend about $800 for new equipment but explains, "This level of communication becomes the lifeblood of your business."

One line, three phone numbers

Richard Ebbs operates two businesses from his home office in Houston, Texas, but manages to get by with only one business line. Ebbs runs a commercial photography business, and also provides marketing consulting to a variety of clients. He is able to juggle business calls by combining two telephone services—distinctive ringing and voice mail. Here's how he does it:

Ebbs signed up for two additional telephone numbers using the Distinctive Ringing service from Southwestern Bell. The first number is printed on his stationery. If Richard is not in, an answering machine takes the call. The second number is programmed to ring his fax modem on the computer. The third number, used as a hotline number to isolate responses to keyed ads, has its own separate answering machine. He uses a Comshare 750 switch that detects the ring tones and sends each call to the appropriate device.

In addition, Ebbs has a Motorola Bravo Express pager connected to voice mail supplied by his paging company. If a caller reaches a busy signal, s(he) is automatically forwarded to his voice mailbox and can leave a message or punch in a phone number. The voice mailbox is programmed to automatically page him.

His phone bill averages $45 a month, and his pager bill averages $15 a month (including pager airtime).

Two lines: one incoming; one outgoing

Dr. Susan Caldwell, a principal in BioInformation Resources, specializes in Internet marketing consulting from her home office in Miami, Florida. The bulk of her work is done online, providing Internet training, marketing research and competitive intelligence for her business clients.

When she started out, she attempted to use one line for business and personal voice calls, as well as for modeming and fax. In less than two months, however, Caldwell decided to install a second line for business. Having a second line eliminated "the psychological stress of knowing that anytime I was on-line, I could be losing business." The business line offers the extra benefit of providing a free *Yellow Pages* listing and a listing in directory assistance.

Now she uses her original phone line for personal calls, modeming, faxing, and outgoing calls, and her new line for incoming business calls. According to Caldwell, the extra cost of a business line was well worth it. "Business has increased considerably and all for only $50 a month," she reports.

Caldwell uses a 14.4 Kbps Supra .32 bis fax modem, which compresses files before sending them. A two-line Panasonic speakerphone handles voice calls. When her business line is busy or she can't get to the phone, calls are forwarded to voice mail. This allows her to record multiple calls that couldn't get through otherwise. The voice mail service, provided by Southern Bell, costs about $16 a month. Says Caldwell, "It's a reasonable tradeoff for small businesses that want to look bigger at relatively little cost."

Two lines, five phone numbers

Bob Mastin owns Aegis Publishing Group, a small publishing house in Newport, Rhode Island that specializes in telecommunications books. Bob provides a variety of ways for customers to reach him using voice lines, answering service, fax and an 800 and a 900 number. Here's how it all works:

Mastin actually has only two physical phone lines in his office, yet he has five different phone numbers. Using Ring Mate (Nynex's version of distinctive ringing), he publishes separate numbers for voice and fax calls, yet runs both over the same phone line. He also has a 900 number that customers call to consult with him and an 800 number used for ordering books. These numbers piggyback onto both lines. He uses a fax switching device (Duophone AFX-200; $120) from Radio Shack, that can recognize the difference between voice and fax calls based on the ring tone.

If line one is busy, or when Bob is unable to answer within two rings, his calls go to an answering service. The service takes a message or an order. The service faxes orders to Mastin's fax machine. Voice messages are also faxed.

Mastin has three Panasonic two-line phones in his facility. One is in his office, another in the main work area, and the third in the mail room. He has a Konica 150 fax machine and a AT&T 1342 digital answering machine on the line that doesn't get forwarded to the answering service.

Mastin pays $4 monthly for Busy Call Forwarding (forwards calls to the answering service) and $6 monthly for Ring Mate. He pays about $100 - $150 a month in long-distance charges, which includes the 800 line, and $60 a month for local phone costs. The answering service costs $80 to $100 a month. The 900 number has a $30 monthly service fee, which is applied against revenues. Since Mastin earns $1.50 for every minute billed at $2.95, he has never had to cut a check for the service fee (in fact, he makes money on the 900 line). Total monthly charges range between $200 and $300. When asked his opinion of his communications costs, he replied, "Considering that I couldn't live without any of my lines, the cost is nominal."

Three lines - key system

When people call their vet, they usually have a very sick pet on their hands. Dr. Gary Hurlbut, who owns the Pacifica Pet Hospital, has a system that keeps him available to his patients, regardless of the time of day. He does this by using a standard phone system during the day, and forwarding his calls to an answering service for coverage after hours and on weekends. If an emergency call comes in, his service pages him. When his Motorola pager beeps, the telephone number of the service is displayed.

Dr. Hurlbut chose an answering service over voice mail because, he says, "Answering services are more friendly and help assure my customers that their call will reach me." Because his answering service knows his business so well, it helps him in many ways. For example: "One night, my service noticed that I wasn't getting any calls. This didn't fit the usual pattern. So they called my office to see if we had remembered to forward our calls. They discovered that we hadn't and notified me at once. Voice mail can't do that."

The hospital connects two phone lines to the AT&T Merlin 410, a key system. If the first line is busy, calls roll over and ring on the second line. One phone is in the reception area; the other in the Doctor's office. If conditions warrant, another two lines can be added.

A third line is connected directly to the Doctor's Supra 14.4 Kbps fax modem. He uses STF OCR software on his Macintosh computer to receive incoming lab reports, which he prints out on his laser printer. If he needs to send a fax that is not already a computerized one, he scans in the document using a Relisys full-page scanner.

Dr. Hurlbut also subscribes to a veterinary database through America Online, with a yearly subscription cost of $200. "The database is definitely worth it," he says. "I can search for the latest scientific articles and have access to specialists who will answer my questions online." Dr. Hurlbut also belongs to CompuServe, which he uses primarily for computer tech support.

His communication costs average around $320 a month (phone bill is about $200; answering service costs $75-$100; paging costs around $5-$6; online costs range from $20 to $30).

Four lines - heavy online use

Barrie Adkins and his wife Deborah operate Banner Investigations, a legal services support firm located in Victoria, British Columbia. They provide international investigations and online research for attorneys, insurance companies and corporate businesses. They specialize in computer-based research, investigating business assets, conducting background checks, and researching acquisitions and mergers.

Starting with just one phone line in 1984, they've added three since, including their lifeline, a modem line for online sleuthing. The other lines are used for voice and cellular.

The Adkins' estimate that they spend 40% of their time online. They subscribe to several services including a British Columbia data service, AT&T's Easylink and CompuServe. Though his firm pays in excess of $14,000 a year in online and telephone fees (one government database alone charges a subscription fee of $500 per year), Barrie advises, "Don't be afraid of these start-up costs. They come back ten-fold."

Banner Investigations markets via targeted e-mail. The firm has recently added an 800 number, good throughout North America. "The 800 number has increased our business twenty percent," Barrie asserts. He estimates that his phone bill ranges from $500 - $700 a month. Though that may sound high to you, Banner Investigations sees it as simply the cost of doing business.

Four lines, three phone numbers

Jeff Sherman operates Warever Computing, a computer consulting business from his home in Los Angeles. He manages his phone lines using a key telephone system (Panasonic KX-T616). Three lines are used for voice; the fourth is reserved for fax and modem calls.

Using Busy Call Forwarding provided by his phone company, Jeff set up a hunt system. Line one is his advertised business line. If that line is busy, calls are forwarded to line two. If line two is busy, calls are forwarded to line three, which also serves as his "home line" that friends and family use to call him.

Sherman added phone company-provided Call Waiting to line three, which "effectively gives me four incoming business lines for the price of three." Line four is connected to a stand-alone fax machine and is also used for outgoing modem calls. In addition to the four lines, Jeff has an 800 number, which is not really a separate line but rings in on his existing numbers.

Warever Computing gets lots of phone bills each month. Sherman pays around $18 a month to his local telephone company for his regional 800 number, another $7 month to his long-distance provider for 800 calls. His first line is a business line (so he can get into *Yellow Pages* and business directory assistance); the other three are residential lines. Business service costs him $26 monthly, which includes Three-way Calling and Call Forwarding. The three residential lines—with Three-way Calling, Call Forwarding, and Call Waiting—cost an additional $60 a month. His long-distance and toll charges add another $40 to $50 a month. Total monthly charges: $150-$160.

Six lines - KSU-less system

Edgewater Productions, located in San Francisco, provides film, video and multimedia services to corporate clients around the world. Charlie Swanson, Edgewater's owner, needs to be available wherever he is. This could be in the cockpit of a plane while shooting a commercial, adjusting lighting in a rented studio, at a script meeting across town, or in his car while searching for the perfect location shot. Here's how he does it:

Swanson runs four standard business lines to his office. Only the first line is advertised in the phone book. The other lines roll over if the first line is busy. If no one is in the office, or all lines are busy, calls go to phone company-provided voice mail.

The lines are connected to eight four-line KSU-less telephone sets, made by GE, that provide two-way intercom, speakerphone, three-party conferencing, do not disturb, line selection and status indicators.

On the road, Swanson carries a PageNet numeric pager that is connected to his voice mail service. The pager service averages around $12 a month. Callers can leave voice messages and request that he be paged. His pager vibrates so that it never interrupts a meeting or destroys a scene. He also has a cellular phone permanently installed in his car. He uses this phone for keeping in touch when out on location. Call Waiting on that line allows him to catch every call.

Swanson pays $250 a month for 300 minutes of cellular time. This is much better than an earlier plan that was costing him between $500 and $600 a month. "It was like watching dollar bills with wings on them, flying away, minute-by-minute," he says. "Now I'm now longer afraid to use the car phone."

Fax is handled by two onboard fax/modems and WinFax PRO software. E-mail comes via America Online.

If you add it all up, Swanson pays about $600 a month for his communications services. But, he explains, "It's a tiny expense for what I'm getting. If I'm not available to clients, I'm out of work."

Seven lines - modular phone system
Oakland Rim and Wheel is run by the brother-and-sister team of Vicky and Justis Fennell. They sell trailer parts to RV and marine trailer shops throughout northern California. They publish a parts catalog and take orders by mail, fax, and phone. As Vicki says, "We're on the phone all day long here." To serve their far-flung customers, they have four standard phone lines, plus one 800 number, two lines reserved for outgoing calls and personal use, and one line shared between a fax machine and credit card service.

They use AT&T's Merlin 410 modular phone system to manage their standard lines. The system lets them share four phone lines among the six stations in their facility and they could expand up to 10 stations. According to co-owner Vicky Fennell, their phone bills are high, averaging around $150 for long-distance and another $400 for local toll calls. Monthly service charges add another $72. They pay an additional $32 a month for a service contract to keep the Merlin in tip-top shape. Monthly total for communications costs: $654.

In a recent cost-cutting measure, they dropped their cellular phone service, which was averaging about $100 monthly. According to Vicky Fennell, "Cellular was an extra convenience but, just cost too much. Maybe we'll go with pagers in the future."

How many lines do you need?
Probably the most frequent mistake small businesses make is relying on too few phone lines. Follow these guidelines to help you size your system:

❑ How many people need telephone access in your organization? A ratio of 1:3 (one line for each three people) is often sufficient. Larger organizations can get by with a much higher proportion of people to lines. A ratio of 1:8 is not unusual. Of course, to achieve such ratios, you need to have some kind of phone system.

❑ How many telephone peripherals (modems, answering devices, fax machines, etc.) will you need to add? Remember, you can only add up to five devices on one line before you run into trouble.

❑ What's your percentage of incoming/outgoing calls? Perhaps you can restrict some stations to incoming only.

❑ What percentage of the day is spent on the phone? How many are voice calls? Data calls? Fax calls? Ask your staff to log all calls for a couple of weeks. Then carefully analyze the results.

❑ What are the opportunities for sharing lines? Can your people share workstations by telecommuting a couple of days a week? Can you use a line-sharing device to switch calls automatically?

❑ Ask your local telephone company to perform a busy line study. You'll have to pay for the study, but the information can be invaluable. You'll get an hour-by-hour look at the frequency and number of busies on your lines.

❑ What are your customers telling you? Survey your clients, vendors and customers to find out how they feel about your accessibility. If they report a lot of busy signals, take heed.

What's next?

Now that you've completed your telecommunications tour, and seen what other businesses are doing, it's time for you to take action. Walk through your facility and inventory your current setup. What's missing? What needs replacing?

Set aside some time to reflect. What can you do to improve customer accessibility? What is your competition doing? What new services could you offer through improved telecommunications? How do you want to position your business to enter the next century?

Prepare your business for the future by taking advantage of some of the telecom technologies outlined in this book. Let telecommunications provide you with your personal on-ramp to tomorrow's I-way. You won't be sorry.

I love getting mail

Send me your questions and comments. Let me know how you handle your telecommunications needs, especially if you are using new technologies in innovative ways. If you're aware of a product or service that you think should be included in the next edition, or have developed an imaginative communications solution, I'd like to hear from you too. Please write in care of:

Aegis Publishing Group, Ltd.
796 Aquidneck Avenue
Newport, Rhode Island 02842

Better yet, send e-mail directly to me at 71022.2131@compuserve.com.

Resources

LAN information

LAN Magazine
600 Harrison St.
San Francisco, CA 94107
800-234-9573

Network Computing
600 Community Dr.
Manhasset, NY 11030
516-562-5000

EDI information

Electronic Commerce magazine
2021 Coolidge St.
Hollywood, FL 33020
954-925-5900

EDI News
1201 Seven Locks Road
Potomac, MD 20854
301-424-3338

EDI information online

CompuServe has an EDI forum. To get there, type GO EDI.

Internet EDI directories:

EDI and Electronic Commerce Resources
(www.disa.org/edi)

EDI Help Desk *(www.wwa.com/unidex/edi)*

Networking software

LapLink
Traveling Software
800-343-8080

OfficeConnect
3Com
800-638-3266

Vines
Banyan
800-2-BANYAN

Netware
Novell
800-NETWARE

LAN Manager
Microsoft
800-426-9400

Pathworks
Digital Equipment Corp.
800-DIGITAL

Windows for Workgroups
Microsoft
800-426-9400

LAN Software Starter Kit
Invisible Software
800-982-2962

LANtastic
Simply LANtastic
Artisoft
800-233-5564

Personal NetWare
NetWare Lite
Novell
800-451-5151

Appendix A
Additional Resources

●●

General books about telecommunications

Winning Communications Strategies: How Small Businesses Master Cutting-Edge Technology to Stay Competitive, Provide Better Service and Make More Money
by Jeffrey Kagan
Aegis, 1997
Contains fascinating real-world profiles that demonstrate how progressive small businesses are benefiting from new communications technology.

Telecom Business Opportunities: The Entrepreneur's Guide to Making Money in the telecommunications Revolution
by Steven Rosenbush
Aegis, 1997
Written by *USA Today* telecom reporter Steve Rosenbush, this guide shows where the money is being made in the evolving, deregulated telecommunications industry.

Phone Company Services: Working Smarter with the Right Telecom Tools
by June Langhoff
Aegis, 1997

Describes phone services in detail, and how to use them in real-life applications.

Telecom Glossary: Understanding Telecommunications Technology
by Marc Robins
Aegis, 1997
Provides a clear overview of the technology and demystifies telecom terms.

Newton's Telecom Dictionary: The Official Glossary of Telecom- munications Acronyms, Terms and Jargon
by Harry Newton
Telecom Library, New York
Harry Newton, publisher of *Teleconnect* Magazine, created this handy compendium of telecommunications terms many years ago and publishes an update frequently.

LAN Times Guide to Telephony
by David D. Bezar
Osborne McGraw-Hill, Berkeley, CA, 1995
Written by telecommunications expert David Bezar, this book is designed for information services professionals.

The McGraw-Hill Telecommunications FactBook: A Readable Guide to Planning and Acquiring Products and Services
by Joseph A. Pecar, Roger J. O'Connor and David A. Garbin
McGraw-Hill Inc., New York, 1993
Full of engineering diagrams, this book gives a high-level overview of telecommunications technologies.

The Telecommuter's Advisor: Working in the Fast Lane
by June Langhoff
Aegis, 1996
A useful guidebook for telecommuters, mobile workers, work-at-homers and remote workers of all kinds.

900 Know-How: How to Succeed With Your Own 900 Number Business, 3rd Edition
by Robert Mastin
Aegis, 1996
The bible of the industry. A nuts-and-bolts start-up guide for toll-collectors on the info highway.

The Business Traveler's Survival Guide: How to Get Work Done While on the Road
by June Langhoff
Aegis, 1997
Learn how to stay connected to home base from virtually anywhere, including international locations. The ideal travel companion.

Magazines

TELECOMMUNICATIONS Magazine
685 Canton Street
Norwood, MA 02062
617-69-9750
www.telecoms-mag.com

Mobile Computing & Communications Magazine
470 Park Avenue South
New York, NY 10016
Subscription Department
PO Box 52406
Boulder, CO 80323-2406
www.mobilecomputing.com

Teleconnect magazine
Telecom Library, Inc., 1995
800-LIBRARY
800-677-3435

Telephone equipment catalogs

Hello Direct
800-444-3556

Radio Shack
800-843-7422

Web sites

Salestar
www.salestar.com
This site has a call pricer that can help you compare long-distance telephone charges and pick a winning plan.

The Telephone History Website
www.cybercom.com/~chuck/phones.html
Contains loads of interesting historical information, sources of old-time telephone equipment, and trivia.

Telecom Information on the Net
www.spp.umich.edu/telecom/telecom-info.html
This is the most comprehensive site on the Internet for telecommunications information. Contains daily updates of telecom and Interent news, links to technical information, FAQs, policy and regulation issues, research, associations, events, and telco companies around the world.

Internet Yellow Pages

Big Yellow (Nynex's national yellow pages)
www.bigyellow.com

International yellow pages
http://wyp.net

AT&T's toll-free directory
www.tollfree.att.net/dir800/

Appendix B
Telephone Companies

....

Major telephone companies - U.S & Canada

AT&T
32 Avenue of the Americas
New York, NY 10013-2412
212-387-5400
www.att.com

Ameritech
30 South Wacker Dr.
Chicago, IL 60606
312-750-5000
www.ameritech.com

Bell Atlantic
1717 Arch St.
Philadelphia, PA 19103
215-963-6000
www.bellatlantic.com

Bell Canada
1000 rue de La Gauchetiere Ouest
Bureau 3700
Montreal, Quebec H3B 4Y7, Canada

514-397-7000
www.bell.ca

BellSouth
1155 Peachtree St., NE
Atlanta, GA 30309
404-249-2000
www.bellsouth.com

GTE
P.O. Box 152092
Irving, TX 75015-2092
214-718-5600
www.gte.com

MCI
1801 Pennsylvania Ave., NW
Washington, DC 20006
202-872-1600
www.mci.com

Nynex
1095 Avenue of the Americas
New York, NY 10036
212-395-2121
www.nynex.com

Southwestern Bell
One Bell Center
Saint Louis, MO 63101
314-235-9800
www.swbell.com

Sprint
2330 Shawnee Mission Pkwy.
Westwood, KS 66205
913-624-3000
www.sprint.com

Stentor
410 Laurier Ave., West Rm. 200
Ottawa, Ontario, Canada K1R 7T3
800-567-7000
www.stentor.ca

US West
7800 East Orchard Road
Englewood, CO 80111
303-793-6500
www.uswest.com

RBOCs

There were seven holding companies created when AT&T was broken up (At this writing, there are six). These companies are called Regional Bell Operating Companies (RBOCs). But they're best known as the Baby Bells. The Baby Bells today are:

- Ameritech
- Bell Atlantic
- BellSouth
- Nynex
- Southwestern Bell (merged with Pacific Telesis)
- US West

LECs

Each Baby Bell owns many of the Local Exchange Carriers (LECs) that provide telephone service to a specific geographic area. For example, Southwestern Bell (SBC) owns LECs that operate phone service in several states, including Alabama, Kentucky, Florida, Georgia, California, and Nevada. But SBC doesn't own all the phone companies in each state. For example, in California, GTE controls a major portion of Southern California and there are several dozen smaller LECs that operate just an exchange or two.

● ●